Self-Assessment for Wastewater Treatment Plant Optimization

Gerard Wheeler
Daryl Burke, M.Sc.E., P.Eng.
Dawn Flancher, P.E.
Barbara Martin

American Water Works Association

Self-Assessment for Wastewater Treatment Plant Optimization

Managing Editor—Books: Melissa Valentine
Cover Art: Melanie Yamamoto
Production: Robert Kern, TIPS Technical Publishing, Inc.

Library of Congress Cataloging-in-Publication Data
Names: Martin, Barbara (Barbara Stricos), 1975- author. | Wheeler, Gerard, author. | Burke, Daryl, author.
Title: Self-assessment for wastewater treatment plant optimization : partnership for clean water / by Barb Martin, Gerard Wheeler, Daryl Burke.
Description: Denver, CO : American Water Works Association, [2017] | Includes bibliographical references and index.
Identifiers: LCCN 2016049816 | ISBN 9781625761903 (alk. paper)
Subjects: LCSH: Sewage disposal plants--Handbooks, manuals, etc. | Sewage--Purification--Handbooks, manuals, etc.
Classification: LCC TD746 .M349 2017 | DDC 628.1/620685--dc23 LC record available at https://lccn.loc.gov/2016049816

Printed in the United States of America

ISBN: 978-1-62576-190-3
eISBN: 978-1-61300-255-1

American Water Works Association

6666 West Quincy Avenue
Denver, CO 80235-3098
303.794.7711
www.awwa.org

Partnership for Clean Water Program Information

With the increasing scarcity of high-quality raw water sources, drinking water utilities may be forced to consider treatment and use of alternative sources, such as impacted surface waters, aquifer recharge and recovery, or reuse. This new reality makes it increasingly difficult to define where "wastewater" ends and where "drinking water" begins, a sentiment echoed throughout the water sector.

Optimizing the performance of wastewater treatment facilities to improve the quality of effluent water discharged to the natural environment can help to protect the quality of surface waters that may be used as a downstream drinking water source. The steps taken to improve treated wastewater effluent quality may also result in improvements in operational efficiencies and operating costs. The Partnership for Clean Water's unique self-assessment process, described in this guide, can help to focus utility optimization efforts and create a culture of excellence.

Established as a parallel program to the successful Partnership for Safe Water, the Partnership for Clean Water's mission is to help utilities preserve environmental water quality and protect public health by optimizing wastewater system operations. Its wastewater treatment plant optimization program focuses on optimizing plant operations to improve effluent water quality, providing a margin of safety beyond current regulatory requirements, and achieving this in as efficient a manner as possible, limiting consumption of natural resources.

This is achieved by completing a comprehensive self-assessment of wastewater treatment plant performance, operations, administration, and energy efficiency; developing and implementing an action plan for improvement; and continuously monitoring progress toward optimization. The self-assessment process represents the core of the program and is designed to be flexible and applicable to wastewater treatment plants of a wide variety of sizes and process configurations, including water reuse facilities.

While any wastewater treatment plant may complete the self-assessment process by following the steps outlined in the full guide, utilities are encouraged to subscribe to the Partnership for Clean Water for access to additional resources, recognition, and other subscriber-only benefits. The Partnership for Clean Water welcomes utilities, worldwide, to participate in this voluntary wastewater utility optimization program and benefit from the program's optimization resources. Learn more about the Partnership for Clean Water and subscribe to the program at www.awwa.org/partnershipforcleanwater.

CONTENTS

ACKNOWLEDGMENTS

The Partnership for Clean Water gratefully acknowledges the contributions made by the primary authors and volunteers who drafted, edited, and provided significant and critical commentary essential to the development of this guide. The Partnership for Clean Water Development Team volunteers, with the support of their organizations, contributed their time and talents to guide program development to ensure the creation of a relevant wastewater optimization program, capturing the components most critical to wastewater optimization at modern treatment facilities. The Partnership for Clean Water also gratefully acknowledges the generous support of American Water Works Association.

Primary Authors

Gerard Wheeler – Global Facilitation, Inc., Oakville, Ontario, Canada
Daryl Burke, M.Sc.E., P.Eng. – EXP Services, Inc., Fredericton, New Brunswick, Canada
Chantal Daigle-Verrier – EXP Services, Inc., Moncton, New Brunswick, Canada
Barbara Martin – American Water Works Association, Denver, Colo.
Dawn Flancher, P.E. – American Water Works Association, Denver, Colo.

Partnership for Clean Water Development Team

Robert C. Cheng, PhD, PE (Chair) – Coachella Valley Water District, Coachella, Calif.
Susheel Arora – Halifax Water, Halifax, Nova Scotia, Canada
Michael Barsotti – Champlain Water District, South Burlington, Vt.
Robert Burton – Pennsylvania American Water Company, Scranton, Penn.
Joseph DeVito – Beaufort Jasper Water & Sewer Authority, Okatie, S.C.
Clayton Edwards – Tulsa Water & Sewer Department, Tulsa, Okla.
Bob Freeman – USEPA, Atlanta, Ga.
Kelly Hagan – Grand River Conservation Authority, Cambridge, Ontario, Canada
Jon Haukaas – Fort Collins Utilities, Fort Collins, Colo.
James Horne – USEPA, Washington, D.C.
Alain Lalumiere – Reseau Environnement, Montréal, Québec, Canada
Aaron Law – Ontario Ministry of the Environment and Climate Change, Toronto, Ontario, Canada
Richard Lieberman – USEPA, Cincinnati, Ohio
Daniel Limacher – City of Calgary, Calgary, Alberta, Canada
Leanne Miller – Water Research Foundation, Denver, Colo.

Russell Plakke – Denver Water, Denver, Colo.
Bruce Macler – USEPA, San Francisco, Calif.
Shawn Rowe – Halifax Water, Halifax, Nova Scotia, Canada
Tom Schippert – American Water Works Association, Denver, Colo.
Michael Sullivan – Oak Creek Water & Sewer Authority, Oak Creek, Wis.

EXECUTIVE SUMMARY

As the Steering Committee Chair for the Partnership for Clean Water (PCW) program during the initial drafting of this essential guide, I am excited about this effort and its potential impacts to the wastewater treatment community. Patterned after the proven Partnership for Safe Water (PSW) program, which has successfully and directly impacted 100 million customers by measurable improvements in their drinking water quality, this program is intended to have a similar positive effect on the optimization of wastewater treatment systems. Although the Milwaukee *Cryptosporidium* outbreak in 1993 served as the driver for the PSW program, no such direct public health correlation exists for the Partnership for Clean Water. Nonetheless, the optimization model created is just as transferable, and relevant, for the optimization of wastewater treatment plants.

Since an early age, we have all been taught the workings of the water cycle (most likely in an elementary school science class), when our instructors drilled into our receptive young minds that no new water is created and that all the water that we are consuming has been recycled. As we grow older, this fact seems to diminish, and the drinking water, wastewater, and recycled water industries seem to becoming increasingly segregated from each other. As an example, the existence of both the Safe Drinking Water Act and the Clean Water Act highlights and demonstrates that these water uses still remain separated in the eyes of the regulatory community. In addition, it is still harder to find conferences that cover water, wastewater, and recycled water in one setting rather than conferences covering each topic individually. Fortunately, there has been recognition in the water industry (encompassing all these uses of water) that if the water cycle exists in nature, then perhaps there should be a nexus among these various uses of water in the industry.

This self-assessment guide provides the methodology for improving wastewater treatment plant performance through a needs assessment that focuses on a few key parameters, including effluent water quality and energy savings. Although numerous other parameters could be included as part of this assessment, a key to the success of the PSW program is the use of a singular parameter, turbidity, as a surrogate parameter for how well the drinking water plant is being operated. I firmly believe that a similar approach will be successful in the wastewater treatment community as well. We hope that you will find this guide to be helpful on your optimization journey, and on behalf of the PCW and the public whom we serve, thank you for being the good stewards of public health and protecting the environment.

Robert C. Cheng, PhD, PE
Partnership for Clean Water – Steering Committee Chair
Assistant General Manager
Coachella Valley Water District
Coachella, Calif.

LIST OF FIGURES

LIST OF TABLES

CHAPTER 1

INTRODUCTION

Why Optimize?

Wastewater utilities face the complex and often challenging task of collecting influent water, which can vary tremendously in its composition, and treating that water to a level suitable for discharge to the natural environment or another intended use. This water may also contribute to a downstream community's drinking water source or be utilized for a specific reuse application. Because of this, wastewater treatment facilities strive to treat their influent water to maintain compliance with all applicable wastewater treatment regulations, proactively protecting environmental water quality and its potential impact on public health, in as efficient manner as possible. Wastewater treatment plant staff must be tenacious in order to achieve these goals, but how can they document that they are accomplishing this very important task?

The wastewater treatment process also represents a significant consumption of power and natural resources for both individual communities and collectively across North America. Wastewater treatment facilities that take steps to minimize power consumption and improve energy efficiency have the potential to reduce operating costs, as well as their consumption of natural resources and environmental footprint. The most advanced wastewater treatment facilities truly operate as water resource recovery facilities, recovering resources such as energy and nutrients for beneficial use. How can these utilities establish a systematic process to continually improve process performance and efficiency?

The simple answer to address both of these questions is to participate in the Partnership for Clean Water (PCW) program. Using the program's practical tools and guidance allows a wastewater utility to document their current performance and evaluate improvements in effluent water quality, while also quantifying improvements in process efficiencies. This is accomplished through a comprehensive self-assessment of wastewater treatment plant operations and performance—a process that is provided in a stepwise and organized fashion throughout this guide. The objective of this process is to identify factors limiting optimized performance so that an action plan for improvement may be developed and implemented. The mission of the Partnership for Clean Water program is to guide wastewater utilities through this self-examination process, using the program's tools and resources. Although the Partnership for Clean Water program was launched in 2016, it was established as a parallel program to the Partnership for Safe Water, a drinking water utility optimization program with over 20 years of history. Partnership for Safe Water subscriber utilities collectively serve a population of more than 100 million across North America. Using similar tools and processes to those provided in this guide, the Partnership for Safe Water program has helped program subscribers to make significant improvements in drinking water quality.

Similar to the Partnership for Safe Water, the Partnership for Clean Water was established "For Utilities, by Utilities" to provide a practical approach to improve operations and effluent water quality in a user-friendly, nonregulatory manner. For utility staff, the objective of this approach is that its application quickly becomes second nature and inherent in day-to-day operations. The program's annual data submission and review process allows a utility to regularly monitor progress while striving to continually improve. Working through the process, the incremental changes and improvements made in process control actually become infectious, and therefore easily transferrable to transform the culture of the entire organization.

The foundation of this voluntary wastewater utility improvement program is the self-assessment process. Through this process, a utility evaluates its present strengths and weaknesses. Each utility can tailor the process to fit its own staff resources and experience based on a schedule set by the utility—there are no time requirements associated with completion of the self-assessment process. The key is to take small, incremental steps to make operational improvements using the practical program tools. The Partnership for Clean Water data collection software allows a utility to collect data and analyze operations very easily, thus allowing fluctuations in daily, weekly, or monthly performance to be trended, reviewed, and improved. Over time, this approach allows the participating utilities to recognize trends and develop prioritized action plans that are managed and scheduled by the utility. This self-assessment process has been the key to the Partnership for Safe Water's success for the last 20 years, and has been proven effective for utilities of any size. The self-assessment process is also the foundation of the Partnership for Clean Water.

The Partnership for Clean Water program is supported by a very talented group of volunteers who provide utility peer review of the submitted Self-Assessment Completion Reports. With this approach, the people who are most familiar with day-to-day utility operations assist participating utilities throughout the entire process. This makes the goal of continuous quality improvement much easier and attainable for utilities of any size.

The Partnership for Clean Water program's goal is to make small, incremental improvements in day-to-day operations that collectively lead toward continuous quality and operational improvements that can improve effluent quality and the efficiency of treatment plant operations. Why not join the best program available to accomplish this critical task?

Wastewater Treatment Terminology

A variety of terms may be used throughout the industry to refer to facilities that treat influent wastewater for discharge into the environment or for another intended use. Some common terms used to refer to these facilities include wastewater treatment plant (WWTP), wastewater treatment facility (WWTF), publicly owned treatment works (POTW), water reclamation plant (WRP), and water resource recovery facility (WRRF). The Partnership for Clean Water acknowledges the use of all of these terms, as they all are used to accurately reflect the nature of the treatment provided at a specific facility. The Partnership for Clean Water also acknowledges the applicability of the self-assessment process, and portions of this guide, to water reuse facilities and applications.

For consistency, however, a single term, wastewater treatment plant (WWTP), was selected as the most accurate reflection of the objective of this guide. The guidance provided relates to the optimization of wastewater treatment plants with regard to increasing the efficiency of the facility as well as improving effluent water quality. While resource recovery is encouraged at facilities for which it is a practical objective, specific guidance for achieving recovery of energy and/or nutrient resources is beyond the scope of this work. Readers are directed toward the wastewater industry resources referenced at the end of this chapter for additional information regarding resource recovery.

An additional term used throughout this guide is 5-day biochemical oxygen demand (BOD_5), which is reflective of the total biochemical oxygen demand present in the water. Plants may analyze and be permitted on BOD_5 or carbonaceous biochemical oxygen demand ($CBOD_5$). The $CBOD_5$ is a measure of the carbonaceous biochemical oxygen demand of the water, accomplished by adding a nitrification inhibitor chemical to the sample prior to analysis, which prevents nitrification from occurring. Although it is difficult to define a relationship that is accurate for all waters, in some cases BOD_5 has been estimated to be approximately 15 percent higher than the $CBOD_5$ concentration. Throughout this book, the term BOD_5 is used to represent both BOD_5 and $CBOD_5$. Plant staff should select the appropriate parameter(s) to consider when completing the self-assessment process. In cases where BOD_5 and $CBOD_5$ must be differentiated, the distinction will be clearly noted in the text.

Partnership for Clean Water Background

The Partnership for Clean Water is a voluntary wastewater utility program that was originally created to parallel the Partnership for Safe Water's successful drinking water optimization program that has been active for more than 20 years.

The Partnership for Clean Water program will ultimately consist of four phases, three of which have been launched at the time of this writing. All program subscribers are encouraged to participate through Phase III, the self-assessment phase, while the optional Phase IV represents fully optimized performance. Participating wastewater treatment plants that complete all four phases of the program are deemed to have achieved fully optimized operations and performance as reviewed, analyzed, and approved by Partnership peer reviewers and optimization experts.

Partnership programs are self-directed and self-paced. They are based on completing a comprehensive self-assessment of treatment plant operations and performance, as described in this guide. Self-assessment completion reports are submitted by treatment plants participating in the program. These reports are peer reviewed by utility professionals and optimization subject matter experts who review reports based on an established process. The reports are evaluated according to specifically defined report evaluation criteria, designed to evaluate a utility's efforts in completing the self-assessment process. The following sections describe the three established phases of the Partnership for Clean Water program as well as key principles of the proposed Phase IV. These phases apply similarly to the Partnership for Clean Water (covered in this guide) and the Partnership for Safe Water drinking water optimization program, which is not covered in this guide. The Partnership for Safe Water program has a dedicated drinking water treatment plant self-assessment guide, available from American Water Works Association (AWWA), titled *Self-Assessment for Water Treatment Plant Optimization*. Additional information about the Partnership for Clean Water and Partnership for Safe Water programs may be obtained by contacting the American Water Works Association. To participate in either program, utilities must apply and meet specific program eligibility requirements.

> Phase I: Phase I is based on utilities making the commitment to participate in the Partnership for Clean Water's wastewater treatment plant optimization program and complete the program through Phase III. Return of the membership application, payment of dues, and provision of access to program materials such as the software, user information, and the published self-assessment guide complete Phase I. At this phase of the program, subscribers are encouraged to review the program resources provided and identify key individuals to be involved in the Partnership process.

> Phase II: Phase II consists of submitting baseline data for selected wastewater treatment parameters, including biochemical oxygen demand (BOD_5), total suspended solids (TSS), total phosphorus (TP), and ammonia nitrogen (NH_3–N), relative to the plant flows and discharge permit levels. Plants are also encouraged to submit baseline data for energy consumption. Wastewater systems use the

program's data collection software to submit 12 months of performance data to the Partnership. After submitting baseline data, plants continue to submit data on an annual basis, which are compiled into an annual report published by the program. By establishing a performance baseline, the utility will be able to quantify the impact of its ongoing optimization efforts on treated wastewater effluent quality. In aggregate, this data also allows the overall impact of the program on treated wastewater effluent quality, and—ultimately—its environmental impact, to be quantified. All data submitted to the Partnership are confidential.

Phase III: Phase III represents the core of the Partnership program and consists of completing the wastewater treatment plant self-assessment process. Subscribing utilities are strongly encouraged to complete Phase III, which is also the basis for receipt of the Directors Award, industry-wide recognition provided by the Partnership for Clean Water for program subscribers that successfully complete the self-assessment process. The self-assessment process is designed to be utility-directed and self-paced, so that it may be completed according to the schedule that provides the greatest amount of information and benefit to utility staff. Utility staff follow the self-assessment process as indicated in this self-assessment guide, in order to identify performance limiting factors (PLFs) and develop a targeted action plan to improve performance. The results of the self-assessment are compiled in the self-assessment completion report, which is submitted to the Partnership for Clean Water for peer review. The Phase III process continues to focus on treated effluent water quality and also includes procedures for evaluating plant and process energy usage, which may then be assessed on an annual basis to quantify the impact of optimization efforts.

The self-assessment completion report (prepared by the utility in this phase) is reviewed by trained utility volunteers, consisting of wastewater treatment plant optimization experts, in accordance with an established process. This group of volunteers is referred to as the Program Effectiveness Assessment Committee (PEAC). The PEAC reviews self-assessment reports according to a set of defined evaluation criteria, which cover the broad categories of Performance, Administration, Operations, Design, and Overall. Reports are peer reviewed to ensure that utilities have made a good faith effort at assessing wastewater treatment plant operations. A successful review results in the utility and treatment plant receiving the program's Directors Award for completion of Phase III of the program. Award-winning utilities receive industry-wide recognition at AWWA's Annual Conference and other venues. Phase III utilities retain their award status by submitting annual progress updates that consist of performance data and a short narrative report that describes optimization activities completed during the annual reporting period. The Partnership recognizes Phase III utilities for maintaining their status and performance level for long-term periods, and utilities are recognized at five-year intervals.

Phase IV: Phase IV is an optional, but encouraged, phase of the program that is planned to recognize wastewater treatment plants that have achieved the highest possible levels of performance. Achieving Phase IV is based on meeting stringent effluent water quality performance goals and demonstrating full optimization. Plants applying for Phase IV program recognition must be Phase III plants in good standing in the program and are required to submit a Phase IV award application, following Partnership for Clean Water guidelines. Phase IV award-winning facilities also receive industry-wide recognition for their achievements. At the time of this writing, Phase IV of the Partnership for Clean Water program is under development.

Partnership for Clean Water and Effective Utility Management (EUM)

The Partnership for Clean Water guidance and self-assessment process is designed to be complementary to and compatible with existing resources and programs for wastewater utility operations and management. One of these resources is AWWA's Utility Management Standards, which includes ANSI/AWWA G510: Wastewater Treatment Plant Operation and Management. This standard includes detailed descriptions of the essential or critical requirements for the effective operation and management of a wastewater treatment plant. Many of these principles are echoed throughout the Partnership for Clean Water's wastewater treatment plant self-assessment process.

Effective Utility Management (EUM) resources are other tools that wastewater utilities may apply to assess and continuously improve utility management practices. EUM is built around *Ten Attributes of Effectively Managed Utilities* and *Five Keys to Management Success*, which covers all aspects of utility operations and is a well-accepted framework to help utilities improve their performance and move toward sustainable operations. EUM resources were developed by a collaborative partnership of several agencies, including the Association of Metropolitan Water Agencies (AMWA), the American Public Works Association (APWA), the American Water Works Association (AWWA), the National Association of Clean Water Agencies (NACWA), the National Association of Water Companies (NAWC), and the United States Environmental Protection Agency (USEPA). Many of these organizations also participate in Partnership programs, such as the Partnership for Clean Water and Partnership for Safe Water. The Ten Attributes include areas such as Operational Optimization, Product Quality, Operational Resiliency, Financial Viability, Employee Leadership and Development, and Stakeholder Understanding and Support. The Effective Utility Management Primer is a descriptive guide that provides a framework for assessing utility management practices according to the Ten Attributes. Areas of the Partnership for Clean Water self-assessment process include consideration of topics that fall within several of the Ten Attributes areas. Completion of the Partnership for Clean Water self-assessment process can be a tool used to support and enhance a wastewater utility's EUM assessment, strategy and progress. Additional information and EUM resources may be obtained online at www.watereum.org.

How to Use This Guide: Self-Assessment Overview

The self-assessment process was developed to assist wastewater treatment plant staff in evaluating current performance with respect to efficiently and consistently producing a compliant effluent water with concentrations of regulated contaminants, such as BOD_5, suspended solids, and nutrients, meeting or exceeding regulatory permit requirements. The self-assessment process also helps utilities identify areas for improvement in energy efficiency, through which potential operational cost savings may be realized. Through completion of the self-assessment process, a wastewater treatment plant evaluates the current level of plant performance with respect to the Partnership's optimization goals and determines if performance improvements are needed. The objectives of the self-assessment are to identify the reasons for less-than-optimized performance by reviewing its operational practices, level of administrative support, and plant unit process capability, and to develop a plan to address the identified limitations in order to improve plant performance. In the process of completing the self-assessment, utility staff may identify and prioritize performance limitations, develop an action plan, and over a period of time, implement improvements to move the treatment plant toward optimized performance. It is important to note that, while some improvements may be implemented to produce near immediate changes, many wastewater treatment process improvements may take a significant amount of time (weeks, months, or even years, in some cases) to realize performance improvements. Both short-term and long-term planning is encouraged in the development of action plans, as will be described in greater detail in chapter 8.

This handbook is a source document for utilities that want to conduct a self-assessment to improve the performance of existing wastewater treatment plants employing a variety of unit processes. The handbook primarily addresses the assessment and optimization of procedures to assist wastewater treatment plants in comfortably meeting typical regulatory permit requirements—for parameters such as BOD, nutrients, and total suspended solids—and achieving these goals in as energy-efficient a manner as possible. However, optimization goals may be developed for any unit process in the treatment plant, and the consideration of these goals is encouraged. Achieving these levels of performance will help to maximize the protection of environmental water quality from contamination by nutrients and other wastewater-related parameters and help to protect source water quality.

It is not anticipated that all utilities will meet the Partnership's suggested optimization goals, or their own internally established goals, when the self-assessment process is first undertaken. One objective of completing the assessment is to better understand and quantify current plant performance, so that areas limiting optimized performance can be identified and an action plan developed that allows the utility to work toward improved performance and, ultimately, optimization.

This guide is not intended to describe cost-saving options or present alternatives for designing new facilities for expansion purposes. The focus of the self-assessment is to maximize the performance of existing facilities by optimizing operations with limited capital expenditures. However, in some cases, the optimization approach described may result in cost savings and/or increased capacity from existing facilities, particularly with regards to improving the energy efficiency of facility operations. Utilities are encouraged to consider correlating their optimization efforts with any potential cost savings that may be achieved. Although beyond the scope of this work, utilities are also encouraged to consider opportunities for resource recovery in applications where it may be practical and beneficial.

This guide follows the framework of the Partnership for Clean Water program; a voluntary program designed to support and recognize utilities that make the commitment to improve environmental water quality by optimizing wastewater treatment plant and collection system operations. Although participation in the Partnership for Clean Water program is not required in order to assess treatment plant operations according to the steps presented in this guide, Partnership participation is encouraged as it provides access to the program's full offering of software tools, support, and resources.

Guide Layout

This guide is designed to be used in conjunction with additional Partnership tools to assist staff in conducting a comprehensive self-assessment of wastewater treatment plant performance and developing a self-assessment completion report. Partnership plants submit the completion report to the program's PEAC for utility peer review and feedback. Successful completion of this process allows plants to be eligible to receive the Phase III Directors Award. Although the guide focuses primarily on the assessment of wastewater treatment plants for nutrient removal and energy efficiency practices, the self-assessment principles and questions may be applied to any type of unit process and/or application, including proprietary processes.

The guide is organized to provide plant staff with steps through the process of completing the self-assessment in a logical fashion. The self-assessment process includes the following categories, each listed with the chapter number in which they are addressed:

- Performance Assessment (chapter 2)
- Unit Process Capacity Assessment (chapter 3)
- Unit Process Performance Assessment (chapter 4)
- Facility Energy Optimization (chapter 5)
- Application of Operational Concepts (chapter 6)
- Administration (chapter 7)
- Identification and Prioritization of Performance Limiting Factors/Action Plan Development (chapter 8)

Additionally, the guide's Appendices contain a variety of resource materials that are relevant to utilities completing the self-assessment process.

Each chapter is organized into three primary sections: Understanding, Status, and Action. The Understanding section provides background information about the topic, the key objectives for the section, selected case study examples, and discussion of the key issues critical to meeting the performance goals. The Status section contains the self-assessment questions to be addressed by plant staff to gain insight into how current plant practices support the performance goals. The Action section contains information that may assist plant staff with developing an action plan to address performance limiting factors and work towards improved performance.

Throughout the guide, references are provided to resources and tools relevant to wastewater utilities that are striving for optimized performance.

A wide range of unit processes are included in this guide, and it is not intended that all processes will be relevant to all treatment plants undertaking the self-assessment process. To complete the Partnership for Clean Water self-assessment process, members of the self-assessment team are only required to answer the self-assessment questions that pertain to the unit processes relevant to their treatment plant.

Composite Correction Program Background

To better understand the self-assessment process, it is helpful to understand the history and original basis for the process, which is described in more detail in this section. The authors encourage readers to refer to the original references if questions arise or if a greater sense of historical context is desired. As older documents are referenced, the reader is reminded that, although the principles and processes associated with completing the self-assessment have changed little in the past decades, significant changes have occurred in the industry. Changes in treatment processes, water quality monitoring technologies, and the regulatory environment may impact specific components referenced in prior documentation.

The format for the self-assessment process was originally derived from the USEPA's work, "Improving POTW Performance Using the Composite Correction Program Approach," published in 1984. This work was written in response to the industry environment at that time—during which many wastewater treatment plants were in need of performance improvements while provided with limited funding to implement such improvements. The work was developed in collaboration with a consultant, Process Applications, Inc. (PAI), which has contributed to the wastewater and water treatment plant optimization effort for more than 30 years. The Composite Correction Program (CCP) approach was subsequently applied to water treatment plants in a work developed in 1991 by the USEPA and PAI. The CCP served as a means to optimize drinking water plant performance for particulate removal to reduce the potential public health risks from pathogens, such as *Cryptosporidium* (Renner et al. 1991). Applied in both wastewater and drinking water, the CCP is a two-part process consisting of an evaluation phase, called a Comprehensive Performance Evaluation (CPE), and a correction phase, called Comprehensive Technical Assistance (CTA). The CCP is a comprehensive, systematic approach for assessing the root causes of performance problems at existing treatment plants. It is comprehensive in identifying the unique combination of factors in the areas of design, operation, maintenance, and administration that are limiting performance. Specific guidance on use of the CCP for achieving optimized performance is provided in USEPA's CCP handbook.

Another resource contributing to the Partnership's drinking water self-assessment content was the Water Research Foundation (WRF, formerly AWWA Research Foundation) workshop titled Optimizing Surface Water Treatment Plant Performance (WRF 1995). During the workshop, approximately 20 water treatment experts from around the United States gathered to discuss water treatment optimization. The workshop participants focused on the factors affecting treatment optimization for particle removal at conventional water treatment plants. The information assembled during the workshop was intended for use by the Partnership for Safe Water's partner organizations (United States Environmental Protection Agency, American Water Works Association, Water Research Foundation, Association of Metropolitan Water Agencies, Association of State Drinking Water Administrators, and National Association of Water Companies) in their development of a utility staff–driven self-assessment processes. Specific criteria for identifying optimized facilities were derived from this workshop (WRF 1997). Both the Partnership for Safe Water and Partnership for Clean Water self-assessment guidance documents stem from the outcome of this workshop, as well as the original USEPA manuals.

The areas in which performance limiting factors have been broadly grouped (administration, maintenance, design, and operation) are all important in that a single factor in any one of these areas can contribute individually to poor performance. However, when implementing a self-assessment program, the relationship of these categories to achieving the goal of consistently and efficiently producing a compliant effluent water must be understood. The relationship of utility administration, design, maintenance, and operation in the production of a compliant treated effluent water on a continuous basis is illustrated graphically in Figure 1-1. As shown, administration, design, and maintenance activities must all work together to result in a plant physically capable of producing a consistently compliant treated effluent. It is the operation, or more specifically, the process control activities that enable a physically capable plant to produce compliant treated effluent on a continuous basis. Figure 1-1 illustrates the most direct approach to improving performance. The starting point is to focus on implementing the necessary process control activities.

For example, if effluent nutrient concentrations cannot be consistently maintained at desired levels because operating staff are not at the plant to make process adjustments in response to changing influent water quality, then improved performance will require better staff coverage. In this case, identified limitations in meeting process control needs (e.g., limitations in making process adjustments) help establish the priority for improving staff coverage (i.e., an administrative issue). Additional staff would alleviate the identified deficiency and allow the processes to be adjusted so that progress can continue toward the performance goal.

Self-Assessment Goals

The Partnership for Clean Water goal-setting philosophy, described in this section, encourages setting optimization goals representative of performance that provides a margin of safety for compliance with regulatory requirements, without incurring extraneous treatment expenses. The program also encourages the development of optimization goals relating to energy efficiency and operational improvements, which also have the potential to reduce plant operating costs.

Plants working on completing the self-assessment are not required to meet these performance goals. However, plants do need to adopt a set of effluent quality and energy efficiency optimization goals in order to assess current plant performance and optimization status as well as complete the Performance Assessment component of the process. It is recommended that plants consider

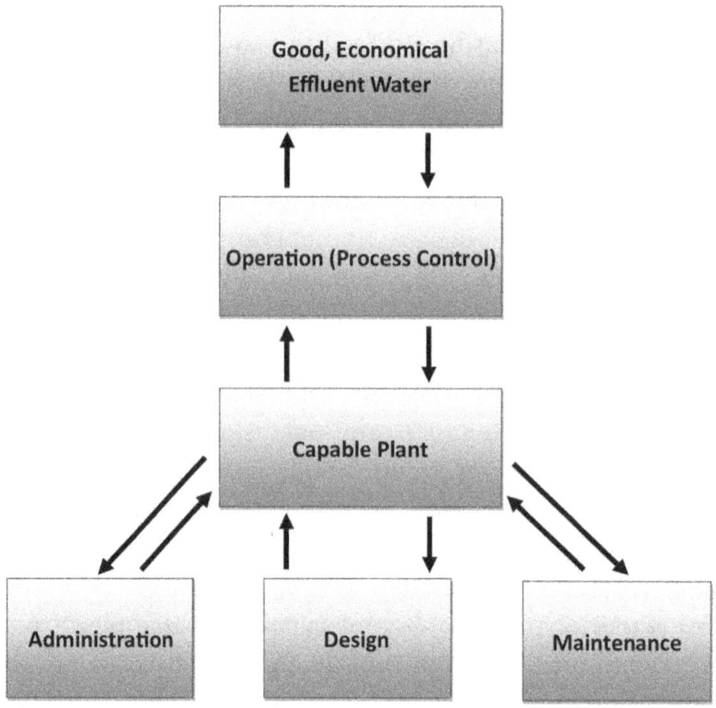

Figure 1-1. Capable plant model

adopting the Partnership optimization goals, as described in the following sections. Wastewater utilities may also consider developing an internal or interim set of treatment plant optimization goals to be used for the assessment and long-term performance planning purposes.

The Partnership's optimization goals for biochemical oxygen demand, total suspended solids, ammonia nitrogen, total phosphorus, and sludge accountability are displayed in Table 1-1. Again, note that these goals are optimization targets and are suggested goals only—not discharge permit regulatory requirements or requirements for successfully completing the self-assessment. Plants can and do successfully complete the self-assessment without attaining the Partnership's optimization goals. Plants may also prefer to develop their own unique set of optimization goals, based on supporting information such as that obtained from special studies, from which to assess performance. An important component of the self-assessment process is the assessment of plant performance relative to specific performance goals.

The effluent quality optimization goals are provided as a percentage of the plant's discharge permit limit. Therefore, the absolute numerical value of the optimization goal for each parameter may vary from plant to plant. If a plant does not have a permit limit for a specific parameter included on this table, they are still encouraged to monitor and report data for this parameter to the Partnership. All data submitted to the Partnership for Clean Water program remains confidential. The data collected and reported as part of the annual data reporting and

Table 1-1. Partnership for Clean Water optimization goals

Parameter	Goal description	Partnership optimization performance goal
Ammonia nitrogen (effluent)	Continuous stable performance regardless of variations in influent and upstream process quality	<95% of plant permit limit
$CBOD_5/BOD_5$ (effluent)		
Total phosphorus (effluent)		
Total suspended solids (effluent)		
Sludge accountability	+/– 15% agreement between theoretical and actual sludge volumes based on use of the activated sludge mass control spreadsheet provided with this guide (refer to chapter 2 for additional detail)	
Energy efficiency (kWh/MG)	Annual reduction in energy consumption per volume of water treated	

self-assessment processes may be evaluated by treatment plant staff for the purposes of establishing baseline performance, the result of which may be used to develop future plant-specific optimization goals. Similarly, plant-specific goals may be set for additional unit processes employed by the plant or for additional water quality parameters, as described in upcoming chapters. If plants do not have a means of monitoring water quality for each unit process described, that should not be considered an obstacle to completing the self-assessment. Rather, plants with gaps in data collection may consider collecting as much data as possible using the means available to them and highlighting areas in which any data deficiencies may exist as potential action items for future improvement (see Table 1-1).

It is important to note that, regardless of the unit process, a major goal for wastewater treatment plants completing the self-assessment is *consistent* treatment plant performance, maintained regardless of changes in influent water quality or flow. Therefore optimization goals should be considered relative to daily/monthly average permit limits, rather than solely the maximum values. A more comprehensive discussion of optimization goals and their relation to the Performance Assessment is included in chapter 2 of this guide.

Getting Started with the Self-Assessment

While utilities may complete the self-assessment in any manner considered to be appropriate, the self-assessment should be a team effort between management and operations, maintenance, and lab staff. The greatest amount of learning and benefit can be obtained by involving staff at all levels in the self-assessment process and making a proactive approach to continuous improvement and optimization a part of everyone's function at the utility, regardless of where a position may reside on the utility's organizational chart. Some suggestions for encouraging broad team involvement in the process include

- Seek management support for completing the self-assessment.
- Formally establish a self-assessment team that includes representatives from across the utility's treatment plant staff. The composition of this team may vary depending on utility size and structure, but it is recommended that the team involve staff from as many levels and functions of the utility as possible.
- Establish regular self-assessment meetings during which staff are encouraged to discuss plant performance as well as specific self-assessment questions. The frequency of these meetings may vary, but many utilities report scheduling team meetings at a frequency of once per week or once per month.
- Act upon team findings: If an action item is identified that is able to, or should be addressed now, don't wait until the self-assessment is completed to take action. Allow staff to take ownership of actions they identify, including conducting special studies or development of operational tools, to start taking steps toward optimization. A report of progress made can be included in the self-assessment completion report.
- Be mindful that the purpose of the self-assessment is not to find faults and assess blame, but rather to serve as a systematic process to identify and correct issues that are negatively impacting plant performance.

The many ways utilities can encourage and promote team involvement in the self-assessment process are limited only by the ideas developed by utility staff. If a utility finds that it cannot complete the self-assessment as a team composed of operations and management staff, it may consider how these attitudes may limit performance, particularly in the administrative section of the self-assessment. The utility may find that working through the self-assessment process as a team results in improved communication and understanding at all levels.

Finally, the self-assessment process is a self-paced process. There is no time limit imposed during which a utility is required to complete the self-assessment. The schedule for completion of the self-assessment is set by the utility in a manner that allows utility staff to derive the greatest benefit from the self-assessment process, while working with the resources available.

After obtaining utility support for the self-assessment process and forming a self-assessment team, the team is ready to develop an approach to completing the self-assessment and begin the process. As discussed previously, there is no deadline for completing the self-assessment, and utilities are free to complete the process according to a flexible timeline that provides the most benefit to utility staff and plant performance. There is no single correct way to complete the self-assessment, although many utilities have implemented regular, recurring meetings of utility staff to discuss the assessment questions and optimization principles.

The structure of this handbook parallels the major steps of the self-assessment process and provides the complete framework required to complete a self-assessment of treatment plant performance according to Partnership for Clean Water guidelines. Partnership for Clean Water subscribers are also provided with program resources and tools designed to support the utility's self-assessment and optimization efforts.

Chapter 1

Chapter 1 provides background information about the self-assessment process and how to begin the process by establishing a team and procedures for completion.

Chapter 2

Chapter 2 describes how to conduct a *performance assessment* to determine the existing level of performance of the utility's wastewater treatment plant. Based on the findings of the performance assessment, the utility staff will know how treatment plant performance compares relative to the Partnership's or the utility's internally established optimization goals. Even if the performance assessment reveals that plant performance is optimized, it is recommended that the utility complete the remaining parts of the self-assessment to more fully examine all aspects of treatment plant operations, including administration, process control, and energy usage. This allows the plant to identify and "polish" any potential factors limiting completely optimized performance, and ensure that the plant's excellent performance is not simply due to a stable and predictable influent quality. Plants that receive a consistently stable and predictable influent stream may derive benefit from the self-assessment to ensure that this predictability does not mask a lack of treatment skills and lead to complacency in plant staff. After completion of the performance assessment, the self-assessment team should then complete chapters 3 through 7. These chapters examine design, operational, and administrative areas to help utilities identify any performance limiting factors that may inhibit optimized plant performance.

Chapter 3

Chapter 3 discusses the *capacity assessment*. An assessment is made of the physical size of the plant's major unit processes (i.e., the concrete basins) to determine if they are of adequate size to meet the desired performance goals to minimize the plant's impact on environmental water quality. These findings will help the assessment team determine if any of the plant's major unit processes are limiting the plant's performance and whether current physical facilities are adequate to meet the optimization goals. If basin sizes are adequate, it is likely that other factors in the areas of maintenance, operation, or administration are the more likely to be limiting optimized performance. Although the objective of the self-assessment is to improve performance through operational improvements while minimizing major capital expenditures, plants that identify a severe design or capacity deficiency using the procedures provided in this chapter may find that more significant modifications to plant design are required to ultimately improve plant performance.

Chapters 4–7

Chapters 4 through 7 outline procedures to identify *unit process–specific*, *operational*, *facility energy efficiency*, and *administrative* factors that may limit performance, respectively. In each chapter, one or more performance limiting factors checklists allow the assessment team to review the material collected during the evaluation of a specific topic (e.g., unit process, operation, or administration).

- ***Chapter 4: Unit Process Performance Assessment.*** Note that the unit process–specific components included in chapter 4 include aspects other than capacity that are not addressed in chapter 3. An example of a design aspect covered in this chapter is the impact that flow distribution between multiple basins has on optimized performance.
- ***Chapter 5: Facility Energy Optimization.*** Chapter 5 is specific to assessing and improving facility energy usage and site-specific energy efficiency (process-related energy efficiency aspects are addressed in chapter 4).
- ***Chapter 6: Application of Operational Concepts.*** The operational topics addressed in chapter 6 include assessment of the plant's process control program and the operational staff's ability to interpret and respond to water quality and process changes with the appropriate controls.
- ***Chapter 7: Administration.*** Chapter 7 addresses administrative issues such as staffing, funding, training, and policy that are required to create a capable plant that allows operational staff to meet the performance goals. Many of the concepts included throughout these chapters are illustrated with utility case studies.

Each chapter contains a list of self-assessment questions for the team's consideration. At the end of each chapter is a table that summarizes responses to each question, including columns to indicate if each factor considered is Optimized, Partially Optimized, or Not Optimized. If the team considers an area to be optimized, the factor listed is not considered to be limiting plant performance and should not be considered further. The self-assessment completion report should contain documentation or an explanation that supports the selected optimization status, which may include information such as that generated by a special study. However, if the status of a particular item is determined to be Partially

Optimized or Not Optimized, the factor is considered to be limiting plant performance and should be considered in the identification and prioritization activities of chapter 8. Likewise, the self-assessment report should include documentation to support the conclusion. Ultimately, factors deemed not optimized should be associated with an action-improvement plan, as described in chapter 8.

Chapter 8

Chapter 8 presents a method for the self-assessment team to identify and prioritize the most important factors that may be limiting plant performance, as determined from the assessment conducted in the previous chapters. This provides clarity to the team as to where follow-up activities should be implemented to improve plant performance. *The goal of the self-assessment is to optimize existing facilities without major capital improvements.* Utilities do not need to wait until the self-assessment is complete to begin addressing action items. Staff may find it beneficial to begin work on action items while the self-assessment is in process. The self-assessment team should document any progress made on action items in the self-assessment completion report (with the understanding that some process improvements may require a substantial amount of time before performance improvements are realized).

It is significant that using the procedures in this guide to assess and optimize performance is valuable for wastewater utilities and treatment plants of all sizes and process configurations. The Partnership for Safe Water program's parallel water treatment plant self-assessment process has been successfully applied at plants of a broad range of sizes and types, with some of the smallest plants to successfully complete the process serving a population as small as 2,000. Over the program's 20-year history, the plants serving the smallest populations were able to improve treated water quality to the same extent as plants serving populations greater than 700,000. While the numerical results of this drinking water program may not be directly transferrable to wastewater treatment plants, the ability of all treatment facilities to improve performance, regardless of size, as a result of staff focus on operations and performance is a concept relevant to both water and wastewater treatment plants alike.

For Partnership for Clean Water subscribers, the optimization process does not end after completion and submission of the self-assessment completion report. The optimization process is ongoing as plant operations, and the plant itself, continue to evolve and change with the passage of time. Once action items are identified and prioritized, utility staff should work on implementing and completing these actions. Utilities should continue to regularly review unit process performance data for the most critical operational parameters to ensure that performance continues to be maintained or improved. Plants that were performing at a very high level prior to completing the self-assessment may not see their improvements reflected as significantly in the data and should look to other areas to measure improvement, such as reliability, consistency, and operator knowledge.

Utilities should also plan to review progress made versus the action plan developed and continue to identify any additional areas that may limit optimized performance. Plants may even want to repeat the completion of the self-assessment process on a regular basis, such as every three to five years, although this is not a Partnership for Clean Water program requirement. If needed, new action plans can be developed to address any newly identified performance limiting factors to ensure that the plant remains on track to continuously improve. By implementing a self-assessment process at a utility, continuous improvement processes can eventually become an embedded component of the utility's culture.

Figure 1-2 presents the recommended steps for completing the self-assessment, as described in the previous section.

Self-Assessment and Its Role in the Partnership for Clean Water

The self-assessment technique is a cornerstone of the Partnership for Clean Water program. The Partnership approach leaves all decision making to the participating utility and provides guidance that allows utility staff to conduct their own specifically tailored approach to achieving optimization. The guidance provided in this self-assessment guide includes questions that prompt utilities for answers that fit their own unique self-assessment process.

Utilities undertake their Phase III self-assessment in a manner that fits the organization and expertise within their specific utility. In most cases, a team is assembled comprised of staff with a stake in optimizing treatment process control or an important role in achieving treatment optimization success. The team may consist of operations, maintenance, instrumentation, laboratory, and administrative/management personnel. In addition, one or two team members typically serve as the "champions" of the self-assessment, guiding the team's efforts and tracking down answers to many of the team's questions. Once the team is assembled, members typically gather data and review the self-assessment guide through a series of meetings. The meetings can be spread over several months or longer. The self-assessment questions are intended to inspire utility practitioners to further search for treatment optimization opportunities. The team may augment their meetings with visits to specific areas of the plant to view the aspects of treatment optimization under discussion.

After completing the self-assessment process with the self-assessment team, using this guide as a tool, the team champion develops a Phase III report that summarizes the self-assessment results. In most cases, the utility will model their report after the Partnership self-assessment completion report template and example reports that are available on the Partnership for Clean Water website. It is important to understand that the report provides information to Partnership utility volunteer reviewers primarily on the quality of the utility's self-assessment

Step 1 Complete Performance Assessment (chapter 2) *Determine current level of plant performance versus optimization goals*
Step 2 Complete Capacity Assessment (chapter 3) *Determine if sizes of major unit processes are limiting performance*
Step 3 Complete Unit Process Performance and Facility Energy Assessments (chapters 4-5) *Identify other aspects of unit process design and energy usage that may be limiting performance*
Step 4 Complete Operation Assessment (chapter 6) *Identify operational practices limiting performance*
Step 5 Complete Administration Assessment (chapter 7) *Identify administration practices limiting performance*
Step 6 Assemble and Prioritize Comprehensive List of Factors Limiting Performance (chapter 8) *Identify activities to address factors that will improve performance*
Step 7 Implement Activities that will Improve Performance
Step 8 Assess Performance Improvements

Figure 1-2. Steps for completion of self-assessment

process and, secondarily, on the quality and improvement in the plant's effluent quality goals.

Utilities have reported significant success in treatment optimization following the Partnership's self-assessment process. The strength of the process is a bottom-up approach to optimization in which the utility undertakes and controls the optimization process using Partnership program guidance and data collection software. Utilities develop and implement action plans that are derived directly from the Partnership's self-assessment process. These action plans allow utilities to strive to identify and greatly reduce complacency and significantly increase reliability.

Plants Undergoing Renovation

Because of the long-term nature of the self-assessment process, some plants may experience a major renovation or process change during this time. Progressive utilities are continuously evaluating current capabilities and improving their facilities to meet future demands and anticipated regulatory requirements. In some situations, treatment plants require major renovations, such as the addition of basins, filters, or completely new treatment processes. These renovations present challenges to both the utility completing the self-assessment and the Partnership for Clean Water self-assessment completion report peer review team. Completion of the self-assessment process can be accomplished while a plant is undergoing major renovations. The following guidance is offered to plants undergoing renovations during the self-assessment process:

- The use of the self-assessment process is recommended as a tool to assist in determining whether the planned renovations adequately address performance limiting factors.
- The utility may consider the Partnership's optimization goals in selecting performance criteria for newly installed processes.
- Treatment plants should consider the timing of the self-assessment completion report submission in relation to completion of the renovation. Construction should be far enough along that specific action plans addressing known performance limiting factors can be included in the submittal package.
- Plants undergoing major renovations are required to provide additional report content. Specific documentation should include operational activities used to optimize the existing plant and an explanation of how the renovation will improve operations, effluent quality, and energy efficiency.
- Include specific action plans describing how the renovation removes existing/previous plant limitations and improves plant operations and effluent quality. For example the self-assessment completion report would contain an explanation of:
    - The potential root cause of the issue that the renovation is intended to address
    - The operational techniques currently implemented to help improve the situation
    - How the renovations will ultimately address the issue
- Plants should include significant future operational changes that may potentially impact effluent water quality. Provide backup documentation used to support the proposed change (consider the use of a special study, described in Appendix B, if the results help to provide support for the modification). Explain the operational and effluent water quality improvements expected after the renovation. If the anticipated improvements cannot be quantified, it may be necessary to postpone completion of the self-assessment until the results of the renovation are available.

The use of pilot study data may be indicated to support the anticipated impact of a plant renovation on operations and effluent water quality, particularly if the pilot study was conducted over an extended period of time and under varying conditions. Although plant renovations can pose an additional challenge when completing the self-assessment process, plants that conduct the self-assessment under these circumstances should be well prepared to meet the challenges of operations when the renovated plant comes online and becomes fully operational.

Self-Assessment Completion Report

Partnership for Clean Water subscribers are encouraged to submit a self-assessment completion report upon completion of the self-assessment process. The report is reviewed by a team consisting of volunteer utility optimization experts known as the Program Effectiveness Assessment Committee (PEAC). Successful peer review is required for plants to become eligible to receive the Partnership's Phase III Directors Award. Utilities that do not participate in the Partnership for Clean Water may also benefit from developing a self-assessment completion report as a means of organizing the major findings and outcomes of the self-assessment process and communicating the information to utility management and other stakeholders.

As there is no single correct way to conduct the self-assessment, there is also no single right way to construct the self-assessment completion report. A report that follows the logical structure of this guide and the steps represented in Figure 1-2 is recommended to most clearly communicate the process undertaken and outcomes observed. A Phase III Completion Report Checklist (defining the minimum acceptable contents for submission of a Phase III self-assessment completion report), a template for creating a Phase III Self-Assessment Report, and an example self-assessment report are available on the Partnership for Clean Water website. Use of the template for Partnership subscribers is strongly encouraged, as the template provides users with a structure from which the self-assessment report can be more easily assembled. The appendices contains examples of additional information that may be included in the report; for example, a special study and example standard operating procedures (SOPs) format.

Staff preparing the completion report may wish to review the appropriate Partnership for Clean Water resources and documentation to ensure that adequate information is provided for the peer review team. After the peer review is completed, utilities will receive feedback regarding the outcome of the evaluation and the team's award determination. Current Partnership for Clean Water subscribers are encouraged to visit the program website (www.awwa.org/partnershipforcleanwater) for the most up to date information regarding the program and report submission requirements.

References

Cheng, Robert, Steve Hubbs, Jim Fay, and Barbara Martin. December 2015. "PSW–20 Years of Progress." *Journal AWWA.*

Effective Utility Management: A Primer for Water and Wastewater Utilities. June 2008. American Public Works Association, American Water Works Association, National Association of Clean Water Agencies, National Association of Water Companies, US Environmental Protection Agency, Water Environment Federation.

Linder, Kevin, and Barbara Martin. 2015. *Self-Assessment for Water Treatment Plant Optimization.* Denver, CO: American Water Works Association.

Renner, R.C., B.A. Hegg, J.H. Bender, and E.M. Bissonette. 1991. *Optimizing Water Treatment Plant Performance Using the Composite Correction Program.* USEPA 625/6-91/027. Cincinnati, OH: USEPA.

USEPA. 1984. *Optimizing POTW Performance Using the Composite Correction Program Approach.* Cincinnati, OH: USEPA.

Water Environment Federation. 2015. *The Nutrient Roadmap.* Washington, DC: WEF.

CHAPTER 2

PERFORMANCE ASSESSMENT

Objective

This chapter pertains to the assessment of both the data collection parameters and the treatment plant's performance with respect to achieving the local regulatory agency's final effluent discharge permit requirements. The performance assessment is designed to understand the performance of the facility relative to the facility's optimization goals. The approach enables WWTP staff to self-assess the performance of their facility and compare the results with the facility's discharge permit requirements and optimization goals. The assessment also provides staff with an understanding of the accuracy of the reported treated effluent quality and plant control data as well as any potential data gaps. The Partnership for Clean Water (PCW) self-assessment completion report requires a minimum of two years of data (baseline year and current year). The assessment questions included in this chapter refer to the most recent 12 months of plant data. *All plants, regardless of size or process configuration, should complete this portion of the self-assessment process.*

Through completion of a performance assessment, utility staff determine if the treatment plant is achieving the level of performance necessary to comply with the treated effluent requirements or plant optimization goals using the existing unit treatment processes and current operational practices. To achieve these levels of performance, a wastewater treatment plant should be able to take a variable raw wastewater influent source and consistently produce a compliant treated wastewater effluent.

Multiple treatment processes such as preliminary treatment, primary clarification, secondary treatment (aeration and clarification), and disinfection are applied sequentially to achieve the desired level of treatment to protect both the environment and public health. Process trains in modern wastewater treatment plants are complex and have many different configurations. To ensure that the performance of each unit process is maximized, the multiple barrier concept is applied. The multiple barrier approach at wastewater treatment plants is illustrated in Figure 2-1. The multiple barrier approach has two key objectives to (1) achieve the required level of treatment via primary clarification and treatment in

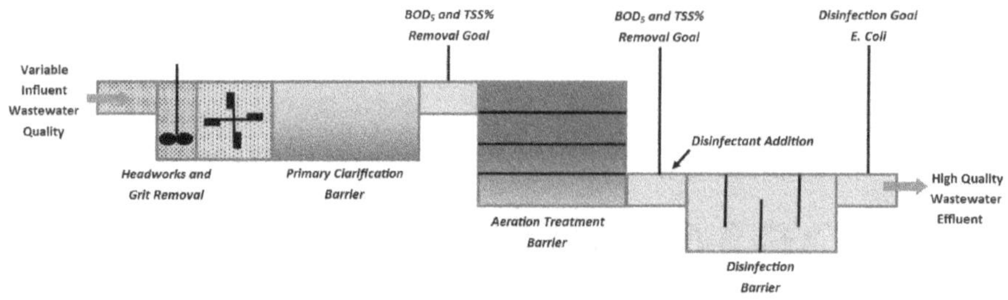

Figure 2-1. Application of multiple barrier approach to wastewater treatment

the aeration basin, to reduce loading and improve performance of the secondary clarification process and (2) achieve the required *Escherichia coli* count via disinfection. Maximizing the performance of the primary clarification barrier reduces the loading to the secondary treatment barrier, resulting in less energy required to treat the BOD_5 in the plant's secondary influent wastewater.

Understanding

The performance assessment determines if specific unit processes consistently achieve optimized performance to ensure the multiple barrier protection is maximized. The performance assessment is based on a comprehensive evaluation of data from plant records, lab reports, process performance data, and process control system information. Typically, a self-assessment team completes the performance assessment, similar to how the remainder of the self-assessment process is recommended to be carried out. Where appropriate, the plant should use online, continuous instrumentation to (1) monitor unit process performance to evaluate treatment performance and effectiveness, and (2) provide real-time information that can be used in daily plant control and operation.

Optimized performance of a treatment plant and its unit processes means consistently achieving a compliant treated wastewater effluent, despite both variations in influent wastewater quality and hydraulic loading changes in plant flow rate associated with rainfall and other events. While the optimization goals will generally include achieving the treated effluent standards, with some margin of safety, the PCW encourages plant staff to cost effectively pursue achieving plant-specific goals that result in a higher level of treatment using the existing unit processes. Refer to chapter 1 and Table 1-1 for a list of the PCW's optimization goals and further discussion pertaining to goal setting.

Setting Goals

What is a goal? A goal is a desired result or a destination. A goal is a specific objective for which to strive. The process of setting goals is important for wastewater treatment plant staff because assessing performance and comparing it with established goals enables utility staff to understand the current status of plant operations. Achieving optimized performance is an iterative process, and assessing performance at set intervals allows the plant to measure progress toward these goals. Goals can be set for any aspect of wastewater treatment plant operation, including individual unit treatment processes. The primary goal of the PCW is to encourage WWTPs to strive for process excellence by meeting or exceeding treated effluent permit requirements in an economical manner and with the existing unit processes.

Utility staff may be successfully achieving the treated effluent standards and elect to pursue an even higher level of treatment with the existing unit processes. This will require adopting new performance goals. These new, plant-specific performance goals should be challenging but considered achievable by the treatment plant. It is important to note that the same goals may not apply to every system, and the amount of time required to achieve the goals may vary. Activated sludge, for example, is a complex biological process. Because of the biological nature of the process, responses to process control changes will take time to emerge and are unique to every WWTP. A rule of thumb for activated sludge systems is that systems typically require four to six Sludge Retention Time (SRT) periods to respond to process control changes. This requires long-term performance goals that are specifically planned to reflect the typical timeline required for a biological system to respond.

A plant can successfully complete the self-assessment phase of the PCW program while still in the process of working toward achieving these goals. The purpose of the self-assessment process is to objectively assess plant performance, identify performance-limiting factors, and develop and implement action plans that will enable the plant to eventually achieve the PCW optimization goals, if this is the desired outcome for the utility.

Staff may prefer to develop their own plant-specific goals while completing the self-assessment process. Setting plant-specific goals should be a process based on thoughtful examination of past data, best practices, and sound justification. If plant-specific goals are used for the self-assessment, an explanation of the goal development process should be submitted as part of the self-assessment completion report.

Plants completing the self-assessment report are required to submit a comprehensive range of data that includes influent and effluent wastewater quality, as well as the mass of sludge produced by plant unit treatment processes.

Setting goals that are challenging, but still attainable, can be an important means of engaging staff and keeping levels of motivation high throughout the self-assessment process. This applies to goal setting for any plant unit treatment process as well as overall plant performance. Although a limited number of plant unit treatment processes are covered in this manual, plants can develop performance goals for any type of unit treatment process applied within the plant. When applying a similar self-assessment process to other types of unit processes, utilities may determine optimization goals based on past performance, industry best practices, theoretically derived performance estimates, or additional parameters.

When completing the self-assessment process, plants will assess their performance relative to the PCW's optimization goals or plant-specific goals, and they will develop and implement an action plan to improve plant performance. Plants may submit a self-assessment completion report while in the process of working to meet the PCW optimization goals; they are not required to meet these goals at the time of submitting the report.

What is Optimized Performance?

Wastewater treatment is complex, and wastewater treatment plants can vary widely in terms of the treatment processes and configurations used at a specific facility. Despite these differences from plant to plant, optimized wastewater treatment plant performance can be defined as consistently achieving the performance goals for each unit treatment process, as well as producing a consistently compliant treated effluent. This approach will ensure that utility staff work to optimize the performance of each individual unit process. Typical industry performance standards describe what each unit process is expected to achieve. Achieving the performance standards for each unit process helps to ensure the multiple barrier concept documented in Figure 2-1 is being maximized.

For example, primary clarifiers typically are designed to achieve 25 to 40 percent removal of biochemical oxygen demand (BOD_5), and 50 to 70 percent removal of total suspended solids (TSS). Figure 2-2 is a graph illustrating the TSS removal efficiency in a primary clarifier. This example plant has two liquid trains, referred to on the graph as PM1 and PM2. These trains are operated in parallel. The primary effluent suspended solids data for PM1 is depicted by the lighter trace. The primary effluent suspended solids data for PM2 is depicted by the darker trace. The 70 percent (high range) TSS removal efficiency is depicted by the upper horizontal line. The 50 percent (low range) TSS efficiency is depicted by the lower horizontal line.

Performance data was evaluated for the period ranging from Oct. 1 to Dec. 23, 2015. During the period evaluated, the average TSS removal efficiency was 63 percent for PM1 and 65 percent for PM2. On average, both primary clarifiers achieved a high level of TSS removal efficiency, and the efficiencies were close to the high range (70 percent) of TSS removal criteria. Consistently achieving the TSS removal criteria demonstrates that the performance of the primary clarifier barrier is optimized at this facility. The data reflects an upward trend, with both primary clarifiers achieving a higher level of removal efficiency in December compared with October.

As demonstrated by this example, trend charts are useful for recording, tracking, and evaluating long-term plant performance, and they can be an excellent means of communicating performance among plant staff and to management.

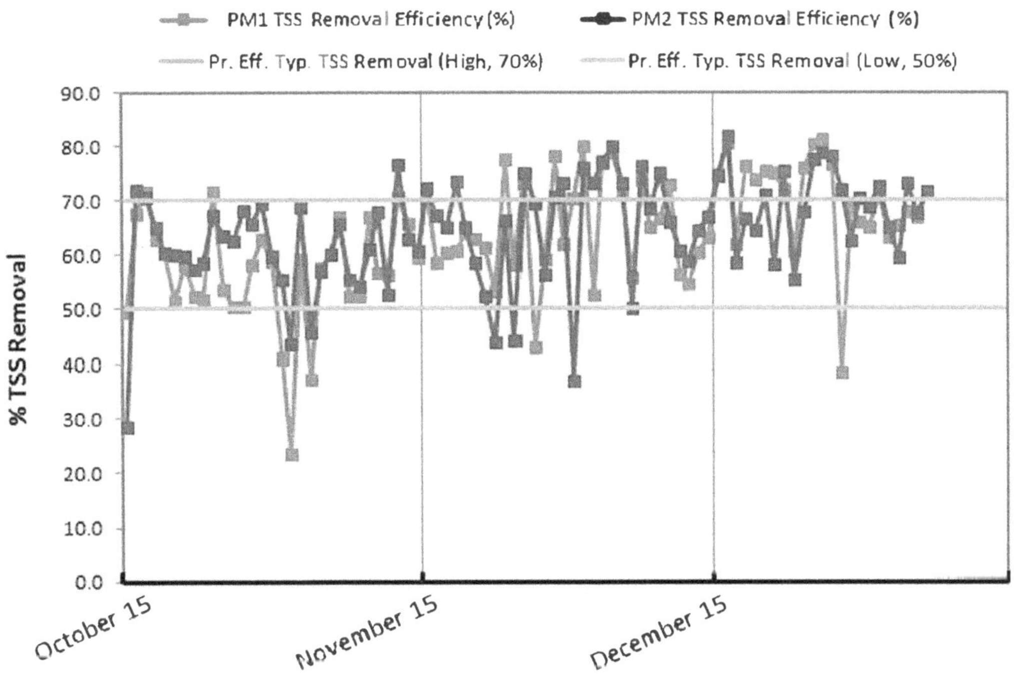

(Source: Gerard Wheeler, Global Facilitation, Inc.)

Figure 2-2. Primary clarifier TSS removal efficiency

Performance Assessment Data Collection

An initial step in the performance assessment is preparing trend graphs for the raw influent wastewater characteristics and final effluent quality. A minimum of 12 months of data are used to evaluate the impacts of seasonal changes in influent wastewater quality and flow rate.

Plants may complete the process using the data collection software tools provided by the PCW or a similar spreadsheet program. Using a spreadsheet program, plant staff may develop similar trend charts to plot additional parameters that provide more information about the performance of the major unit processes.

Table 2-1 outlines the recommended minimum data collection requirements for successful completion of the self-assessment process.

Once the data collection has been completed, the self-assessment team should evaluate the data to determine if sufficient data is being collected and if the treated effluent permit requirements or the plant-specific goals developed by utility staff are being consistently achieved. The self-assessment team should also review the raw influent wastewater and final effluent quality trend graphs to look for any anomalies or inconsistencies in the data. Often, a WWTP may be achieving the desired final effluent quality on an average basis; however, administrative or operational practices may cause the final effluent quality to fluctuate. An example of this is provided in Figure 2-3.

Table 2-1. Self-assessment minimum data collection requirements

Raw influent wastewater flow rate
Monthly/Daily Average (mgd or m^3/d) Month Peak Day (mgd or m^3/d) Design Flow Rate (mgd or m^3/d) Peak Hourly Flow Rate (gpm or L/s)*
Raw influent wastewater characteristics
Monthly Average and Daily Maximum BOD_5 (mg/L) Monthly Average and Daily Maximum TSS (mg/L) Monthly Average and Daily Maximum Total Phosphorus (TP) (mg/L P) Monthly Average and Daily Maximum Total Kjeldahl Nitrogen (TKN) (mg/L N)
Final effluent quality
Monthly Average and Daily Maximum BOD_5 (mg/L) Monthly Average and Daily Maximum TSS (mg/L) Monthly Average and Daily Maximum TP (mg/L P) Monthly Average and Daily Maximum Total Ammonia (mg/L N)
Treated effluent permit requirements
Monthly Average and Daily Maximum BOD_5 (mg/L) Monthly Average and Daily Maximum TSS (mg/L) Monthly Average and Daily Maximum TP (mg/L P) Monthly Average and Daily Maximum Total Ammonia (mg/L N)

Note: Plants are encouraged to report any additional plant-specific and/or unique permits in the self-assessment completion report.

* In many cases the raw wastewater pumping equipment and plant headworks (i.e., screening and grit removal) is upstream of any equalization/attenuation tank(s) and is often sized to handle peak hourly flow rates.

Figure 2-3 shows a graph of the final effluent ammonia concentration from a wastewater treatment plant that will be required to nitrify in the future. The two-season final effluent ammonia limits are depicted by the line on the graph, which fluctuates between 2 and 5 mg/L on a seasonal basis. The two liquid trains in use at the facility are designated on the graph as Plant 1 and Plant 2. The final effluent ammonia concentration for the Plant 1 liquid train is depicted by the dotted trace, while the Plant 2 liquid train is depicted by the darker trace. The effluent ammonia monthly average data evaluated on this graph covers the time period from September 2009 to May 2012. This time period includes three winter periods, which are considered to run from January to March. In this case, the data cover the years 2010, 2011, and 2012. Winter periods are of particular concern because colder wastewater temperatures cause the rate of nitrification to decrease. When colder wastewater temperatures are experienced, any weaknesses in the nitrification process control program will be exposed, resulting in higher effluent ammonia concentrations.

The graph illustrates considerable variability in the rate of nitrification during the winter months in 2010 and 2011, compared with the same period in 2012. Between January and March 2010, the final effluent ammonia concentrations from both plants ranged from 8 to 10 mg/L, clearly higher than the final

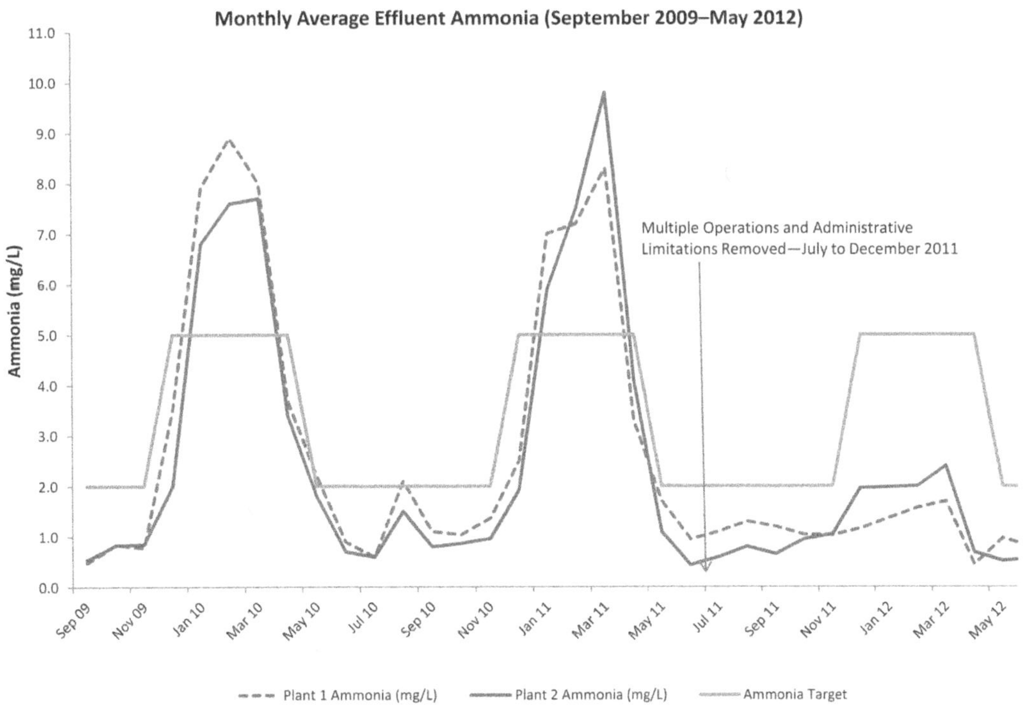

(Source: Gerard Wheeler, Global Facilitation, Inc.)

Figure 2-3. Comparison of Plant 1 and Plant 2 final effluent ammonia concentrations

effluent ammonia target of 5 mg/L for this time period. The same pattern was repeated between January and March 2011. Utility staff responded by identifying and resolving multiple operational and administrative performance limiting factors, starting in July 2011. A higher and more scientific level of process control was established, including calculating the quantity of sludge mass required to support consistent nitrification. A new process control program was implemented in July 2011. This allowed adequate time to learn and stabilize the new process control program in advance of the approaching 2012 cold winter season.

In the example, as a result of the new program implementation, the operating staff successfully maintained stable nitrification between January and March 2012. In addition, the level of nitrification achieved was significantly higher compared with the two previous winters. During that same time period, the final effluent ammonia concentrations for Plants 1 and 2 were both below 2 mg/L, and well below the seasonal final effluent ammonia limit of 5 mg/L. The level of nitrification achieved by the operating staff was unprecedented, and demonstrates how resolving administrative and operational performance limiting factors can result in a significant improvement in performance (these topics are addressed in greater detail in chapters 6 and 7, respectively).

Variations in the stability of the activated sludge barrier could indicate possible problems with plant operations (see chapter 6, Application of Operational

Concepts), such as an inadequate amount of activated sludge required to achieve consistent nitrification or limitations in the major unit processes capacity (see chapter 3, Unit Process Capacity Assessment). Understanding issues, such as the reason for the erratic performance of the nitrification process, is important. This type of initial finding can be used to focus activities for the remainder of the self-assessment. For example, when reviewing the process control program, the self-assessment team may evaluate if the quantity of activated sludge required to achieve consistent nitrification is accurately calculated. Even if the self-assessment team concludes that the mass of activated sludge is adequate to achieve consistent nitrification, there may be a problem with how it is distributed between the aeration basins and secondary clarifiers. Nitrification occurs in the aeration basins, so the majority of the activated sludge should be in the aeration basins, with a minor amount in the secondary clarifiers. The self-assessment team should focus on measuring the distribution of the activated sludge between the aeration basins and secondary clarifiers. The sludge mass control software tool provided to PCW subscribers can aid in quantifying sludge mass for this process.

Compromised performance of even one aspect of the process control required to achieve nitrification can have the potential to impair the treated effluent quality. Completing the self-assessment process, identifying performance limiting factors, and developing and implementing action plans to allow the utility to achieve these goals demonstrates the ethics and tenacity of plant staff for consistently producing a compliant treated wastewater effluent that meets the plant's optimization goals and protects both the environment and public health.

Analyzing trend charts is an important activity when evaluating unit process performance data. These profiles are a valuable means of visualizing the performance of any unit process, which allows plant staff to quickly identify negative trends and implement the correct process control adjustment to respond to changes in performance. Additional examples of unit process trend charts are included in chapter 4.

Sludge Accountability Analysis

An important self-assessment tool that enables utility staff to have confidence in the accuracy of the reported final effluent data is a sludge accountability analysis. The function of a sludge accountability analysis is to calculate the theoretical quantity of sludge that an activated sludge treatment plant is projected to produce. The projected quantity of sludge is compared with the actual measured amount of sludge produced, using process control data and plant records. Typical sludge production rates for similar processes are used to calculate the projected sludge mass produced over the same period of time. Table 2-2 summarizes the data required to complete a sludge accountability analysis. The table comprises two components: the measured quantity of sludge mass produced (shown in

Table 2-2. Produced compared with projected measured sludge mass

	Sludge Mass kg/d
Measured final effluent sludge to receiver	229 (1)
Measured primary sludge transferred to digesters	8,279 (2)
Total measured sludge produced	8,508 (3) = (1 + 2)
Projected primary sludge	**3,649 (4)**
Projected biological sludge	**4,247 (5)**
Projected chemical sludge	**93 (6)**
Total projected sludge	**7,990 (7) = (4 + 5 + 6)**
Sludge accountability	**((7 – 3)/(7)) × 100 = –6.5%**

Roman type) compared with the projected quantity of sludge produced (shown in boldface type).

Interpretation of Results

Referring to Table 2-2, plant records measured a total of 8,509 kg of sludge produced per day. The theoretical quantity of sludge projected to be produced using typical industry sludge production criteria was 7,990 kg per day. More sludge was measured than what was projected. The difference between the theoretical and actual values was –6.5 percent. A sludge accountability of +/–15 percent is generally considered to be acceptable. The example analysis was within the +/–15 percent range that is considered acceptable. The cause of the discrepancy between theoretical and actual vales is typically a combination of minor errors associated with either the flow rate measurement or the concentration of solids measured and reported. The sludge accountability results confirm that staff can have confidence the final treated effluent parameters reported in Table 2-1 are most likely accurate.

If the result of the sludge accountability analysis was outside the acceptable +/–15 percent range, questions relating to the accuracy of the data will arise that need to be answered. These results would draw into question additional data generated for the plant, and would require follow-up action items to identify the factors preventing the sludge accountability analysis from being within the acceptable +/–15 percent range.

Status

These self-assessment questions should be completed for all plants completing the self-assessment process, regardless of process configuration. Throughout the guide, if a particular question is not applicable, plant staff should respond "not applicable" with a brief explanation describing why the question is not applicable.

To determine the plant performance status, develop and review the trend graphs for raw influent wastewater, primary clarification, secondary treatment, tertiary treatment (if applicable), and disinfection for a minimum one-year period. Utility staff are encouraged to evaluate data from multiple years, which may aid in the identification of seasonal or temporal trends. The evaluation of performance data from any additional unit process that the assessment team may choose to evaluate can also aid in determining overall plant status. Based on this information, review the following items:

- Does the final effluent quality meet the PCW (Table 1-1) or plant-specific performance goals?
- Do the primary clarifiers achieve 25 to 40 percent removal of BOD_5?
- Do the primary clarifiers achieve 50 to 70 percent removal of total suspended solids (TSS)?
    - ▲ Figure 2-2 provides an example illustrating achievement of 50 to 70 percent TSS removal through primary clarification.
- If the plant has multiple process trains, does each train perform equally, or are there significant differences in the performance of individual process trains?

If the utility has set additional performance assessment goals for other unit processes in the plant or other parameters that influence PCW or internal performance goals, the following questions should also be considered in relation to these processes/parameters, in addition to the questions above. Utility staff should be aware of all potential influences on unit process performance. For example, do changes in raw influent wastewater quality or flow rate impact the performance of unit processes, such as primary clarifiers, etc.?

- Do changes in wastewater flow rate, implemented to perform routine operational or maintenance practices, or influent quality impact the performance of any unit treatment process?
    - ▲ Planning to implement operational or maintenance procedures should include consideration of the impact on hydraulic loading to the other major unit processes.
    - ▲ Sludge removal practices and schedules should take into account hydraulic loading on units remaining in service. Historical raw wastewater organic and solids loading can be used to predict the impact while the unit(s) are out of service.
    - ▲ Flow meters should be calibrated and maintained to ensure accurate and reliable performance. Meter calibration and maintenance activities should be recorded and documented. Accurate flow measurement is an important aspect to ensure that a representative sludge accountability analysis is completed.

- Do all of the unit processes meet their performance goals, or does treated effluent quality degrade despite consistent and stable process control?
  - Did assessment of the performance of individual liquid trains reveal that some units did not achieve the same performance as the other units in service?
  - The process control program should include a routine assessment of the performance of each unit process in each liquid train. When inconsistent performance is found, the unit should be removed from service and the cause of the performance problem identified and corrected.
  - Special studies should be conducted on the individual unit processes in each liquid train to attempt to identify the most probable cause of poor performance.
- Are the treated wastewater effluent permit requirements being achieved even though other upstream unit processes do not meet the specified optimization goals?
  - Even if the treated effluent standards are being achieved, that does not negate the value of proper operation and stable performance of all unit processes in the plant. This situation illustrates the importance of adopting a multiple barrier approach to protect both the environment and public health.
- Are there any positive or negative trends in the data that should be addressed or further evaluated?
  - For example, Figure 2-2 shows a noticeable upward trend in the TSS removal efficiency over time. Trends such as these should be evaluated to determine the root cause and whether additional action may be required to address such performance trends.
  - Trend charts of raw influent wastewater, primary clarifier effluent, secondary and tertiary treatment, and disinfection, such as those generated by the PCW data collection software, should be developed and evaluated to provide information about influent quality variability and consistency of performance of the major unit treatment processes.
  - Trend charts of raw wastewater influent parameters can indicate variability of the source. The greater the variability, the greater the process control required to achieve optimized performance on a continuous basis.
  - Utilities that have a consistent, high quality raw wastewater must guard against complacency and ensure that plant staff have procedures in place to respond to infrequent and unplanned variations in raw wastewater influent quality.

  - ▲ The potential impacts that changes in raw wastewater quality could have on unit processes, such as primary clarification, secondary and tertiary treatment, and disinfection, need to be understood and controlled so that treated wastewater effluent quality and unit process performance is not impacted.
  - ▲ The potential impact of recycle streams (from digester supernatant and solids dewatering) on treated wastewater quality should be understood. This concept is discussed further in chapter 4.
- Is the activated sludge barrier adequate to achieve the performance goals and scientifically calculated based on flow rate and organic loading?
- Does the actual sludge mass produced agree with the projected sludge mass produced to within +/–15 percent?

Action

If problems are identified during the status review of the plant's performance assessment that prevent the plant from achieving the desired levels of performance, action plans should be developed, scheduled, and initiated to mitigate these problems. This section offers some alternatives to consider for addressing identified deficiencies and performance limiting factors.

- If the results of the performance assessment indicate that the plant consistently produces a high quality treated wastewater effluent, and the individual unit processes are optimized, the performance of the plant is excellent. Any activities identified in the remaining areas of the self-assessment should reflect further refinement of operational or administrative practices rather than the need to make significant improvements.
- Even if the utility meets the performance goals, an assessment should be completed to ensure that the consistent performance is not related solely to having a stable raw influent wastewater loading. Low hydraulic and organic loading rates can mean that the operators' process control capabilities are rarely challenged, and it can be important to address potential complacency issues in this situation.
  - ▲ Treatment of a consistent, raw wastewater with low loading rates can lead to operator and administrator complacency. Consequently, treatment performance may suffer during infrequent events, such as storm flows or industrial discharges that alter the raw wastewater quality, which may challenge the staff's process control capabilities.
- If the plant is unable to produce consistent treated wastewater effluent quality and/or the performance of individual unit processes are erratic, consider:

- ▲ Possible deficiencies with the process control program, process chemistry, or application of process control concepts (such as an inadequate activated sludge barrier, see chapter 6, Application of Operational Concepts).
- ▲ Possible insufficient capacity of the plant's major unit processes (e.g., primary clarification, secondary treatment, tertiary treatment, and disinfection (see chapter 3, Capacity Assessment).
- ▲ Possible impact of poor recycle water quality or inconsistent recycle flows (see chapter 4, Unit Process Performance Assessment).
- ▲ Investigation of process control problems within any of the unit processes that may be causing poor performance. Do not ignore maintenance items in this investigation.
  - ▪ For example, if the aeration basins are adequately designed and meeting their performance goals, but the secondary effluent ammonia concentration is consistently higher or erratic, assess the possibility that low air flow rate to one of the aeration basins is preventing the dissolved oxygen residuals from being high enough to achieve complete nitrification (see chapter 3, Capacity Assessment and chapter 4, Unit Process Performance Assessment).

- If the major unit processes are meeting their performance goals but the treated wastewater quality shows elevated total phosphorus (TP) concentrations, determine the effectiveness of any chemical coagulants (e.g., iron or alum salts) applied to remove TP (see chapter 4, Unit Process Performance Assessment).
  - ▲ Evaluate the type and quantity of coagulants used to remove TP from the treated wastewater.
  - ▲ The chemical sludge produced from applying chemical coagulant will influence the data input to the sludge accountability analysis.
  - ▲ The quantity of chemical sludge produced depends on whether iron or alum salts are applied to remove TP.
  - ▲ Depending on the choice of coagulant applied to remove TP from the final effluent, operations staff should calculate how much chemical sludge is produced, and adjust the activated sludge barrier in the aeration basin to ensure an adequate barrier of activated sludge is maintained (see chapter 4, Unit Process Performance Assessment).

Performance Limiting Factors Summary

This section summarizes the results of the performance assessment to compare plant performance with optimization goals, based on the review of the monitoring data. In Table 2-3, staff would indicate whether performance is Optimized,

Partially Optimized, or Not Optimized. Development of action plans should be considered in areas for which a response of Partially Optimized or Not Optimized is selected.

Regardless of the performance assessment status, all plants should complete the remainder of the self-assessment process to identify additional opportunities for improving plant performance, reliability, operations, or administration. Even the best performing wastewater treatment plants have the potential to improve by undertaking the self-assessment process. Well-performing plants should complete the self-assessment process to help ensure that optimized performance is not achieved due to factors such as low influent organic loading rates or conservative unit process design. In these situations, the self-assessment process can be beneficial toward helping operators avoid complacency and maintain and that wastewater treatment skills.

References

Hegg, B.A. 1999. *Upgrading Existing Secondary Clarifiers To Enhance Process Controllability To Support Nitrification*. Regional Municipality of Halton.

Hegg, B.A., L.D. Demers, and J.B. Barber. July 1989. *Handbook: Retrofitting POTWs*. USEPA 625/6-89/020. Cincinnati, OH: USEPA, Center for Environmental Research Information.

Hegg, B.A., L.D. Demers, J. H. Bender, and E.M. Bissonette. October 1998. Handbook: *Optimizing Water Treatment Plant Performance Using the Composite Correction Program, 1998 Edition*. USEPA 625/6-91/027. Cincinnati, OH: USEPA, Center for Environmental Research Information, Technology Transfer and Support Division, National Risk Management Research Laboratory, Office of Research and Development.

Ontario Ministry of Environment and Energy. January 1992. *Assessment of Factors Affecting the Performance of Ontario Sewage Treatment Facilities*. Ontario Ministry of Environment and Energy: Environment Canada.

Ontario Ministry of Environment and Energy. January 1994. *Assessment of the Comprehensive Performance Evaluation Technique for Ontario Sewage Treatment Plants*. Ontario Ministry of Environment and Energy.

Ontario Ministry of Environment and Energy. July 1995. *Assessment of the Comprehensive Technical Assistance Technique for Ontario Sewage Treatment Plants*. Ontario Ministry of Environment and Energy.

Wheeler, G.P., C.D. Walsh, and B.A. Hegg. 2010. "How to Tap the Human Infrastructure to Gain Cost Effective Performance and Capacity Benefits at a Wastewater Treatment Plant." Paper presented at WEFTEC 2010, New Orleans, October 5.

Table 2-3. Performance assessment questions

Self-assessment category	Questions for gauging whether system is optimized	Response (choose one)		
		Optimized and Documented	Partially Optimized	Not Optimized
Performance assessment	Did the final effluent quality meet the PCW (Table 1-1) or plant-specific performance goals?			
	Do the primary clarifiers achieve 25 to 40 percent removal of BOD_5?			
	Do the primary clarifiers achieve 50 to 70 percent removal of total suspended solids (TSS)?			
	If the plant has multiple process trains, does each train perform equally, or are there significant differences in the performance of individual process trains?			
	Do changes in wastewater flow rate, implemented to perform routine operational or maintenance practices, or influent quality impact the performance of any unit treatment process?			
	Do all of the unit processes meet their performance goals or does treated effluent quality degrade despite consistent and stable process control?			
	Are the treated wastewater effluent permit requirements being achieved even though other upstream unit processes do not meet the specified optimization goals?			
	Are there any positive or negative trends in the data that should be addressed or further evaluated?			
	Is the activated sludge barrier adequate to achieve the performance goals and scientifically calculated based on flow rate and organic loading?			
	Does the actual sludge mass produced agree with the projected sludge mass produced to within +/–15 percent?			

CHAPTER 3

UNIT PROCESS CAPACITY ASSESSMENT

The focus of this chapter is to determine if the existing key unit treatment processes are of adequate capacity to handle the plant's average flows while still meeting the performance goals. Chapter 4, Unit Process Performance Assessment, focuses on other aspects of physical facilities, such as process flexibility and chemical feed facilities, which affect the ability of the plant to achieve the desired performance goals. Plants with sufficient unit process capacity to treat average flow rates should have the capability of making operational improvements that can optimize performance. In some cases, plants that lack capability primarily due to inefficient unit process capacity may find that operational improvements are helpful but may have limited overall impact. In these cases, capital investment to increase unit process capacity may be necessary to significantly improve performance. *This portion of the self-assessment process is applicable to all unit process configurations and should be completed by all plants.*

Understanding

In an evaluation of a wastewater treatment plant, the potential of existing unit treatment processes to achieve the desired levels of performance must be established. If major unit processes have adequate capacity but plant performance is poor, that poor performance is most likely caused by other issues related to the maintenance, administration, or operation of the facility. Normally, factors in these areas can be corrected through low-cost, nonconstruction, operational alternatives. If the major unit processes have inadequate physical capacity to handle the plant's average flows, utility owners should consider modification of these processes as the initial focus for achieving desired performance. These options will be discussed in greater detail in the Action section of this chapter. The overall self-assessment process is invaluable for all utilities to perform. Within the overall self-assessment process, *the primary objective of the major unit-process capacity evaluation is to determine if plant unit processes are adequately sized and whether performance improvement is likely to be achieved solely due to operational improvements.*

Each major unit process is evaluated to determine if its physical size is able to meet required performance goals at the annual average flow rate. This type of capacity evaluation may be carried out for all types of unit treatment processes. The potential capacity of a unit process should not be lowered if minor modifications, such as providing additional chemical feeders or installing baffles, could be accomplished by the plant staff. This approach is consistent with the self-assessment intent of assessing the adequacy of *existing* facilities to determine the potential of non-construction, operational alternatives to achieve optimized performance. It is the experience of the Partnership program that most plants evaluated could achieve a consistently compliant effluent quality without major capital improvements required. In fact, it is anticipated that the steps taken to optimize performance may actually result in operational cost savings for many wastewater treatment facilities.

The major unit process capacity assessment is very beneficial to both older plants, where changes in average flows and loading rates may have occurred over the years, as well as newly designed plants.

Performance Potential Graph Approach

Major unit processes can be efficiently and effectively evaluated through the review of process design calculations or through the development of a performance potential graph. A performance potential graph allows users to project the capacity of each unit process compared with annual average flow rate, using typical design criteria. Such a graph can be completed manually, using a computer-based spreadsheet program, based on the procedure outlined in the CCP handbook (Renner et al. 1991) and the example calculations shown throughout this chapter and available from wastewater design references.

As an initial step in preparing a performance potential graph, the evaluator is required to use judgment to rate the treatment capability for each of the plant's major unit processes. It is important to note that the ratings are based on achieving performance goals for primary treatment, secondary treatment, tertiary treatment (if applicable), and disinfection processes such that *each unit process* achieves the desired level of performance. This allows the plant's entire treatment train to provide *multiple barriers* for the removal of organics, solids, and nutrients from the influent water. The projected treatment capacity of each major unit process is then compared to the average annual flow rate treated during the most recent 12 months of operation. If the most recent 12 months are not indicative of typical plant flow rates, the evaluator may choose to review a time period considered to be more representative.

Average annual flow is defined as the average flow rate treated by the plant during the previous year.

To compare project treatment capacity for comparison with the plant's annual average flow rate, use a performance potential graph as shown in Figure 3-1. The graph provides a visual representation of the relationship between

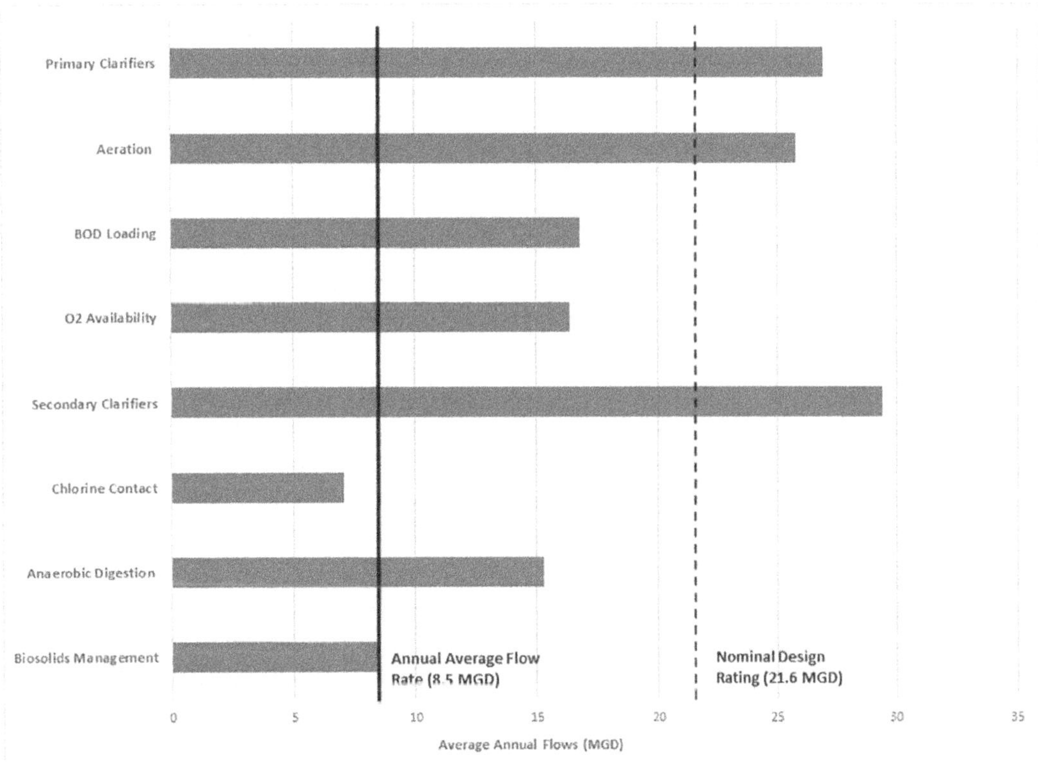

(Source: Gerard Wheeler, Global Facilitation, Inc.)

Figure 3-1. Example process performance graph

the capacity of the processes evaluated and the plant's annual average flow rate. The lengths of the bars on the graph represent the projected capacity of each unit process. The two vertical lines across the graph represent the actual annual average flow rate (left, solid line) and the nominal design flow rate of the facility (right, dashed line).

The performance potential graph provides a visual illustration of the capacity of the major unit treatment processes compared with the annual average flow rate and nominal design capacity. A unit process may be labeled as Type 1, 2, or 3 depending on the capacity of each unit process compared with the actual annual average flow rate, as presented in Figure 3-1. Definitions for Type 1, 2, and 3 are provided in Table 3-1.

Options are available that may allow some unit processes that are marginally meeting annual average flow capacity requirements to meet the plant's performance goals. Often, operational changes can function as work-arounds for perceived physical limitations without making major capital improvements. In some cases, this can be achieved with increased process control or with the use of alternative chemicals where applicable.

Note that in Figure 3-1, the capacity of all unit processes is considered Type 1 and capable of meeting the plant's annual average flow rates, with the exception of disinfection (chlorine contact). The capacity of the disinfection process is 83

Table 3-1. Unit process capacity ratings

Type	Description	Implications
1	Process capacity is adequate; rated capacity exceeds 100% of annual average flow rate.	Major construction should not be required to correct any performance limitations.
2	Process capacity is marginal; rated capacity is between 90% and 100% of annual average flow rate.	Deficiencies may be solved with workarounds. Major construction is often not required. Improved process control and/or elimination of other factors may allow the plant to meet performance goals.
3	Process capacity is too small to meet performance goals; rated capacity is less than 90% of annual average flow rate.	Major construction may be required to improve performance.

Note: Wastewater collection systems are beyond the scope of this guide but anticipated to be incorporated into future PCW resources.

percent of the annual average flow rate and is rated as Type 3 due to its undersized basin capacity with respect to annual average flow. Although disinfection is rated as incapable according to this design evaluation, operations staff should still endeavor to achieve compliance for *E. coli*. Disinfection is achieved through a combination of chlorine dosage and hydraulic retention time (HRT). By varying the chlorine dose, the operations staff may be able to compensate for the relatively short HRT in the contact basin. The general procedure for assessing the capacity of processes not included in this guide is described later in this chapter.

Rating Individual Unit Processes

Many regulatory agencies have adopted wastewater treatment plant design criteria with recommended loading rates for unit processes. General wastewater treatment plant design criteria may be accessed in a variety of sources, including *Recommended Standards for Wastewater Facilities* (commonly referred to as Ten States Standards throughout this chapter) (Health Research, Inc. 2014), *Wastewater Engineering: Treatment and Resource Recovery* (Metcalf & Eddy, Inc. 2013), and *Water and Wastewater Calculations Manual* (Lin 2001). These guidelines, as well as any local regulatory requirements, should be considered in conjunction with the suggested values quoted in this section.

The key activity in preparing a performance potential graph is selecting appropriate loading rates for each major unit process, as needed. Once the loading rates are selected, the capacity of a unit process can be projected. The criteria presented in Tables 3-2, 3-3, and 3-4 are shown to help plants select loading rates for individual unit processes. A wide range of values are included in the criteria, which can translate into large differences in the projected unit process capabilities. Plant staff may also refer to design specifications to assist in the capacity assessment, which may be particularly applicable for newer facilities.

In addition to the information provided in this chapter, evaluators should not overlook their personal experiences or forsake their own judgment in projecting a unit process capability. For plants using a proprietary unit process in their treatment train, staff may need to consult the equipment manufacturer for specific recommendations regarding the proprietary process. Specific guidance to enhance the approach for assessing each unit process is described in the following sections.

Preliminary Handling

Equalization Basin

The equalization basin provides an area for storage of incoming flows in order to reduce flow fluctuations entering the plant, as well as the potential impact of flow fluctuations on downstream unit treatment processes. This may include dedicated tanks at the wastewater treatment plant, the use of otherwise unused tanks at the wastewater treatment facility, as well as storage capacity available in the collection system under some circumstances. Equalization basins are typically sized based upon the cumulative flow anticipated over a 24-hour period, with a buffer of up to 50 percent excess capacity to account for extreme situations.

The volume of storage available in equalization basins is calculated by determining the total volume of all equalization basins used at the plant:

$$\text{Equalization Basin Total Storage Volume (gal)} = \\ [\text{Number of Basins} \times \text{Basin Volume (gal)}] + \text{Additional Storage Volume (gal)}$$

The basin volume is calculated by applying the appropriate volume equation for the shape of the tank. To compare this volume with the plant's flow rate, divide the volume in gallons by 1,000,000 to obtain a total storage volume in units of million gallons (MG).

To assess the capacity of the equalization basin, relative to a 24-hour holding time target, compare the amount of influent flow received by the plant in a 24-hour period, add in the appropriate amount of buffer, and compare this to the basin volume. For example, a plant receiving an average flow of 5 MGD has a total equalization basin volume of 1 million gallons. In this case, the equalization basin represents a holding time of just under five hours, which falls short of the plant's 24-hour holding time goal. The plant may choose to physically increase its storage capacity or identify other strategies for increasing the holding time benefit that the equalization basin can provide.

The equalization basin capacity is compared to the plant's storage time/volume goals to determine whether equalization is a Type 1, 2, or 3 process.

Primary Clarification

The wastewater treatment plant's primary clarification process physically removes solids from the influent water by settling or flotation. Primary clarification is one of the most basic and widely applied forms of wastewater treatment. Following the multibarrier approach to wastewater treatment, an effective primary clarification process can help to reduce solids loading on the downstream treatment processes. Typical design criteria for primary clarifier tanks may be accessed in a variety of wastewater treatment plant design resources, as indicated in the References section at the end of this chapter.

Primary clarification capacity is assessed by selecting a surface overflow rate (gpm/ft^2) relative to the basin depth that will result in acceptable particle removal in the basin. The surface overflow rate (SOR) that is selected should take into account the depth of the basin, whether primary clarification is followed by secondary treatment, and whether waste activated sludge is returned to the primary clarifier. Greater liquid depths generally result in more quiescent conditions that allow for the use of higher SORs. Ten States Standards recommends a minimum side water depth of 10 feet in primary clarification basins. If sludge is manually removed from the clarification basins, adequate depth is required to allow sufficient volume for sludge storage. Sludge removal is a critical component of the proper operation of a primary clarifier. Accommodations need to be made within the process design to ensure that sludge removal is performed at an adequate frequency. Even basins equipped with sludge collection equipment can have difficulty removing solids, depending on specific basin design. Refer to the Sludge Handling section of this chapter for additional information regarding sludge removal pump capacity.

Selecting an ideal surface overflow rate for the determination of primary clarifier capacity is dependent on local conditions. An example of typical primary clarifier surface overflow rates is provided in Table 3-2. Surface overflow rates may be plant and process specific. For example, particles may vary in their properties, resulting in differences in settling velocities, which can impact the ideal SOR for the process. Additional factors that can influence the ideal SOR for a primary settling process include:

- The treatment objective for the primary clarification process in terms of BOD_5 and/or solids removal.
- Influent quality and particle characteristics.
- Use of chemicals to aid in the settling process.
- Water temperature.

Particles with high settling velocities will settle in a basin at relatively high surface overflow rates, while particles with slow settling velocities may not settle effectively, even at relatively low overflow rates.

Primary clarification capability must be assessed based on local conditions. Basins that appear overloaded may have the potential to be optimized by the addition of a chemical treatment to improve particle settling characteristics, or by actions such as increasing the frequency of sludge removal to increase the effective basin depth. The use of baffles may also aid in optimizing the dispersal of basin flow.

Table 3-2 shows criteria suggested for selecting a surface overflow rate to be used in development of the rated capacity for primary clarification. Note that local regulatory requirements may vary with regards to primary clarifier surface overflow rates.

The criteria shown may vary depending on local conditions. Surface overflow rates for manufacturer-specific (i.e., proprietary) units are typically available from the manufacturer and should provide adequate settling for the specific process. For more recently designed plants, the design SOR for the primary settling basins may also be obtained from the plant design specifications.

The capacity of the primary clarification process, with respect to the peak instantaneous flow, may be calculated by multiplying the SOR by the total surface area of all basins and converting it to MGD for comparison with the plant's average or peak instantaneous flow. Although the overflow areas of the weirs and launders may be taken into consideration for this determination, they are not included in the example provided below. Maximum weir loading rates are provided in Ten States Standards and other design resources documents.

$$\text{Capacity of Primary Clarification (gpd)} = \\ \text{Total Surface Area of Basins (ft}^2\text{)} \times \text{SOR (gpd/ft}^2\text{)}$$

This value is then converted to MGD for comparison with the plant's average flow rate to ensure that the primary clarifier basins are of sufficient capacity to meet flow requirements and to determine whether primary treatment is categorized as a Type 1, 2, or 3 treatment process.

However, basin size is not the only variable driving optimization of the primary clarification process. Several other factors may impact process capacity and effectiveness. The self-assessment questions included in the following sections of this chapter relate to physical aspects of the secondary clarification process that

Table 3-2. Primary clarifier: general surface overflow rates

Primary clarifier characteristics	Design average flow (gpd/ft²)	Design peak hourly flow (gpd/ft²)
Tanks not receiving activated sludge	1,000	1,500–2,000
Tanks that receive activated sludge	700	1,200

Sources: Wastewater Committee of the Great Lakes–Upper Mississippi River Board of State and Provincial Public Health and Environmental Managers. 2014. Lin, Shar Dar. 2001.

may potentially limit its capacity for optimization. Chapter 4 contains additional self-assessment questions that address operational aspects of the process.

Secondary Treatment

Secondary wastewater treatment consists of the biological treatment processes applied for the removal of suspended solids, BOD_5, and nutrients. The primary processes for capacity assessment in the secondary treatment processes include the activated sludge process (aeration basin) and the secondary clarifiers. In addition to basin size, capacity assessment of the secondary treatment process should include an evaluation of return activated sludge (RAS) and waste activated sludge (WAS) pumping capacity to ensure that pumps are sized appropriately to meet process operational requirements, as well as an evaluation of aeration equipment to confirm aeration capacity is sufficient to maintain adequate dissolved oxygen (DO) concentrations in the aeration basin. Aeration basin and secondary clarifier capacity may be plotted on the plant's performance potential graph as part of the capacity assessment process.

Numerous factors impact the effectiveness of the activated sludge process in achieving its treatment objectives. The evauation carried out in this chapter is primarily to assess if physical factors, such as basin size, limit the potential for process optimization. The self-assessment questions contained in this chapter refer primarily to physical factors (basin size, pumping capacity, solids removal capacity, and aeration capacity) that may limit optimized performance. Achievement of the plant's treatment objectives is highly dependent on optimized operation of all plant unit treatment processes, including secondary treatment processes. Self-assessment questions pertaining to the operation of all plant unit processes are included in chapter 4.

Activated Sludge

Many factors can potentially impact the activated sludge process, such as BOD_5 loading rates, food/microorganism (F/M) ratio, solids concentration, and hydraulic detention time. This section focuses on the physical aspects of the process necessary to assess capacity. A variety of design references, including Ten States Standards, recommend the following process loadings for activated sludge processes, as displayed in Table 3-3.

Provided that loadings are within the recommended ranges, as indicated in Table 3-3, the hydraulic retention time (HRT) in a conventional aeration basin typically ranges from four to six hours, with sufficient dissolved oxygen (DO) supply provided. Hydraulic retention time varies according to the type of activated sludge process employed. While conventional activated sludge processes may have an optimal hydraulic retention time of four to eight hours, the optimal HRT for extended aeration processes is 20 to 35 hours, and sequencing batch reactors may vary from 15 to 40 hours. As discussed in chapter 4, a concentration goal of 2.0 mg/L DO is applicable to many conventional aeration basin processes.

Table 3-3. Activated sludge: recommended process loadings

Process type	Organic loading (BOD_5)	Food/microorganism (F/M) ratio	Mixed liquor suspended solids (MLSS)
Conventional step aeration, complete mix	40 lb/d × 1,000 ft^3	0.2–0.5 lb BOD_5/d per lb MLVSS	1,000–3,000 mg/L
Contact stabilization	50 lb/d × 1,000 ft^3	0.2–0.6 lb BOD_5/d per lb MLVSS	1,000–3,000 mg/L
Extended aeration single stage nitrification	15 lb/d × 1,000 ft^3	0.05–0.1 lb BOD_5/d per lb MLVSS	3,000–5,000 mg/L

Sources: Wastewater Committee of the Great Lakes–Upper Mississippi River Board of State and Provincial Public Health and Environmental Managers. 2014. Lin, Shar Dar. 2001.

According to Ten States Standards design guidelines, the water depth in the aeration basin most typically ranges from 10 to 30 feet.

To calculate the physical capacity of the aeration basins relative to the plant's average flow and optimal HRT for the process, calculate the total liquid volume of the aeration tanks relative to the worst-case HRT. From this, the capacity of the basins with respect to the plant's average or peak instantaneous flow may be estimated.

$$\text{Total Tank Volume (ft}^3\text{)} = \text{Number of Tanks} \times [\text{Length (ft)} \times \text{Width (ft)} \times \text{Depth (ft)}]$$

Note that if aeration basin size varies per basin, calculate the volume of each basin independently and add them together. Convert the tank volume in cubic feet to volume in gallons by dividing the volume in cubic feet by 7.48. If desired, the volume can then be expressed in MG by dividing the value obtained by 1,000,000. Then compare the total tank volume to the HRT to determine the capacity of the basins.

$$\text{Capacity (gpd)} = [\text{Total Tank Volume (gal)} / \text{HRT (hours)}] \times 24 \text{ (hours/day)}$$

Divide the calculated capacity by 1,000,000 to obtain capacity in MGD, and compare this with the plant's annual average flow to ensure that the aeration basins are of sufficient capacity to meet average flow requirements. The total capacity of the aeration basins is then plotted on the plant's performance potential graph or compared with the desired flow rate to determine whether the aeration basin is categorized as a Type 1, 2, or 3 treatment process.

Once the capacity assessment has been completed based on basin size, the remaining self-assessment questions in this chapter focus on other physical aspects of the activated sludge process that may limit its ability for optimization.

Secondary Clarification

Similar to primary clarification, the secondary clarification process physically separates solids from the liquid process stream located downstream of the activated sludge process. Secondary clarifiers are optimally designed to allow sufficient area for the settling of mixed liquor suspended solids (MLSS), along with effective solids removal and recycling to the aeration basin through the use of the return activated sludge (RAS) process. Specific design parameters may be obtained from a variety of sources.

Ten States Standards provides general design recommendations for secondary clarification basins, including minimum water depth, which is recommended at 12 feet for secondary clarifiers downstream of activated sludge processes and 10 feet for secondary clarifiers located downstream of attached growth processes, under most typical conditions. The recommended surface overflow rate for secondary clarifiers receiving activated sludge are as follows in Table 3-4.

Surface overflow rates may also be determined through pilot testing. The surface overflow rate used in the capacity determination may be taken from standard design references (if the actual design surface overflow rate is unknown), pilot data, the plant's specific design recommendations, or the equipment manufacturer, particularly for equipment using a proprietary design.

To assess the physical capacity of the secondary clarification basins with respect to the plant's annual average flow, (1) calculate the capacity of the basins using the selected SOR from the information provided previously, and (2) convert the calculated capacity to MGD. Although the overflow area of weirs and launders may be considered in this determination, this has not been included in the example equation that follows. Maximum loading rates for weirs can be found in Ten States Standards and in other wastewater treatment plant design references.

$$\text{Capacity of Secondary Clarification (gpd)} = \text{Total Surface Area of Basins (ft}^2\text{)} \times \text{SOR (gpd/ft}^2\text{)}$$

The capacity of secondary clarification (gpd) value is then compared with the plant's average flow to ensure that the secondary clarifier basins are of sufficient

Table 3-4. Recommended surface overflow rates: secondary clarifier

Process type	NTE surface overflow rate (gpd/ft²)
Conventional activated sludge	1,200
Extended aeration, single stage nitrification	1,000
Two stage nitrification	800
Activated sludge with chemical addition for phosphorus removal to <1.0 mg/L P	900
Tanks following trickling filters or RBCs	1,200

Sources: Wastewater Committee of the Great Lakes–Upper Mississippi River Board of State and Provincial Public Health and Environmental Managers. 2014. Lin, Shar Dar. 2001.

capacity to meet the plant's flow requirements. The total capacity of the clarifier basins may then be plotted on the plant's performance potential graph or compared to the desired flow rate to determine whether secondary clarification is categorized as a Type 1, 2, or 3 treatment process.

Basin size is not the only variable driving optimization of the secondary clarification process. Several other factors may impact process capacity and effectiveness. The self-assessment questions included in the following sections of this chapter relate to physical aspects of the secondary clarification process that may potentially limit its capacity for optimization. Chapter 4 contains additional self-assessment questions that address operational aspects of the process.

Attached Growth Biological Treatment Processes

The attached growth biological treatment processes include trickling filters and rotating biological contactors.

Trickling filters. Trickling filters are a fixed-film biological reactor that use media (rock or plastic) over which wastewater is distributed on a continuous basis. Treatment occurs as wastewater trickles over the attached biofilm. The effluent is collected in an underdrain that leads to a sedimentation basin, where solids are further separated from the liquid stream. Trickling filters are a common, low-energy process for secondary treatment. They have better sludge thickening properties, create a non-bulking sludge in the clarifiers, and produce no issues with MLSS control and sludge wasting. However, when compared to a conventional activated sludge process, trickling filter performance is poorer. Trickling filter effluent TSS and BOD removals are lower, and the process is more sensitive to lower temperatures, odors, and sloughing events. It is difficult to accomplish nitrogen and/or phosphorus removals. Trickling filters are often used in combination with activated sludge processes to take advantage of the best of both processes: energy savings and effluent quality. The capacity of trickling filter processes should also be assessed as part of the capacity assessment.

Rotating biological contactors. Rotating biological contactors (RBCs) are a modification to the attached growth process. Plastic media allows for higher loading rates and improved operations. In the RBC process, plastic media (in the form of closely spaced circular disks) are rotated in the wastewater treatment tank as opposed to pumping wastewater to distribute over the media in a trickling filter configuration. Several of these units are typically installed in a series to accomplish treatment needs. They are simple to operate and low in energy costs. The capacity of RBCs should also be assessed as part of the capacity assessment.

Suspended Growth Biological Treatment Processes: Lagoons

Lagoons are shallow earthen basins with mechanical mixing or aerators to provide oxygen for biological treatment and to keep biological solids in suspension.

Principal types of suspended growth lagoon processes are facultative, aerobic flow-through, and aerobic with solids recycling. BOD removals are dependent on SRT and can range from three to six days for domestic wastewater.

Sludge Handling

Optimized operation of the wastewater treatment process is dependent on having the proper mass of sludge distributed correctly among the plant's basins, as well as having the ability to remove accumulated solids from basins (such as the primary clarifiers) when appropriate. The plant must also have the ability to move appropriate volumes of sludge in order to effectively waste sludge and return sludge to the aeration basin as required. Due to the critical nature of the solids handling process, some degree of redundancy is recommended. Plant staff should refer to any local regulatory requirements that may exist for equipment redundancy. The sludge mass control concept, introduced in chapter 2 and discussed in greater detail in chapter 4, provides a strategy for determining sludge distribution between the aeration basin and secondary clarifiers in order to optimize unit process performance.

At the wastewater treatment plant, sludge is collected and moved by using sludge handling equipment and pumps. Since the ability of this equipment to handle sludge can directly impact the treatment capacity of the plant, sludge handling equipment should be considered in the capacity assessment. To assess the capacity of sludge handling equipment, plant staff should consider the following self-assessment questions:

- Do pumps have sufficient capacity rating to handle the total solids volume, including if the largest pump is out of service?
  - ▲ Review the pump capacity ratings of each pump, relative to the required solids volumes to determine if a deficiency exists.
- Is pump efficiency routinely tested to quantify performance?
  - ▲ Pump performance can significantly degrade over time, impacting the pump efficiency.
  - ▲ Pump performance may be tracked over time to proactively manage pump maintenance, rehabilitation, and replacement activities.
- Are any maintenance issues limiting the performance of sludge handling/removal systems in use at the plant?

The capacity and effectiveness of sludge handling equipment is also dependent on maintaining the proper concentration of solids that are pumped. If the concentration of RAS, WAS, or any solids flow is not optimized, this may impact the capacity of the pumps to effectively transport and distribute solids, even if the theoretical pump capacity appears to be adequate.

Advanced/Tertiary Treatment

A variety of advanced/tertiary treatment processes exist for wastewater treatment and reuse applications. A comprehensive discussion of all tertiary treatment processes is beyond the scope of this guide. The tertiary treatment processes in chapter 4, Unit Process Performance Assessment, include tertiary filtration (media filtration) and membrane bioreactors (MBR). These are also the specific processes referenced for consideration in the capacity assessment. Plant staff are encouraged to assess the capacity of additional, relevant tertiary treatment processes employed at their facilities, by applying similar techniques to those included in this chapter, to determine if the physical capacity of any tertiary treatment process is a factor limiting optimized performance of the treatment plant.

Filtration

The tertiary filtration process allows suspended solids to be removed from the effluent water to a higher degree than can be achieved through secondary clarification alone. Tertiary filtration is often used in chemical phosphorus removal applications. To achieve the wastewater treatment plant's effluent goals, it is important that filters are performing adequately on a consistent basis. The design and capacity of the filters are two factors that contribute to the ability to optimize the filtration process. The guidelines discussed in this section are appropriate for media filters. Plants using granular activated carbon filters for the removal of specific organic contaminants may also consider empty bed contact time (EBCT) in their capacity determinations. The capacity of the filtration process is based on a filter loading rate of gpm/ft^2. As with other key treatment processes, redundancy in filtration equipment is preferred, with many local regulatory agencies requiring some degree of redundancy in tertiary filter design, such as the ability to provide full treatment capacity with one filter out of service.

The filter loading rate for wastewater treatment plant tertiary filtration typically ranges from 3–4 gpm/ft^2, with the Ten States Standards recommendations not to exceed a filter loading rate of 5 gpm/ft^2. Selecting a plant's optimal filter loading rate should take into consideration local conditions, applicable regulatory requirements, plant design specifications, and demonstrated filter performance. The typical quality of the water applied to the filter can also impact the optimal filter loading rate. For example, different media types (such as mono-media sand, granular activated carbon, dual or mixed-media, mono-media anthracite), media depths, effective sizes, and uniformity coefficients can effectively handle different hydraulic loading rates and produce similar filtered water quality.

Filtration capacity ratings may be restricted to certain maximum values because of existing local regulatory requirements. In these cases, following local regulatory requirements may be necessary to project filter treatment capability. Data from special studies and/or pilot work may also help plant staff to determine the plant's optimal filter loading rate.

At the end of the filter run, tertiary filters require backwashing to clean the media and restore treatment capacity. Ensuring that filter backwash is properly conducted is critical to optimizing filter operation. Along with assessing the capacity of the filters, it is important to evaluate the backwash equipment and controls for any potential limitations that exist. Limitations in backwash facilities, or inadequate backwash facilities or controls, can negatively impact filter capacity. This potentially requires filter loading rates to be estimated in the more conservative portion of the range required for wastewater treatment, which may be considered a factor limiting optimized plant performance. However, addressing limitations relating to backwash may often be achieved through minor modifications, and therefore is considered to have a less significant impact on capacity than the hydraulic loading rate.

The physical capacity of the plant's filters may be calculated by multiplying the total surface area of the filter in square feet by the optimal filter loading rate. Typical wastewater treatment plant surface loading rates range from 3–5 gpm/ft^2. The plant may also use known design specifications and/or results of special studies or pilot work for the determination of filter capacity. An example of the filter capacity calculation is provided below:

$$\text{Filter Capacity (MGD)} = [(\text{number of filters}) \times (\text{filter area } (ft^2))] \times [\text{filter loading rate } (gpm/ft^2)] \times 0.0014$$

The calculated capacity of the filters is then compared to the plant's average flow rate to evaluate the capacity of the process and determine whether it is considered to be a Type 1, 2, or 3 process. This evaluation may also be performed with one filter assumed out of service to evaluate the degree of redundancy available at the plant.

Membrane Bioreactors (MBR)

Membrane biological reactors (MBR) consist of a bioreactor with suspended biomass and solids separation via microfiltration membranes. An MBR offers one unit process for wastewater treatment, replacing the need for secondary clarification and tertiary filtration. They operate at higher MLSS concentrations and hydraulic loading rates, and they have shorter reactor retention times. MBRs produce a high-quality effluent in terms of low turbidity, TSS, and BOD. High capital and operation and maintenance (O&M) costs, and membrane fouling, cleaning, and replacement, can be disadvantages to the process.

As with the other plant unit processes discussed in this chapter, membrane bioreactors should be sized to perform adequately relatively to the plant's average flow rates, as rated by the manufacturer and allowed by regulation. Membrane bioreactors that have adequate capacity to meet annual average flow requirements are considered to have adequate capacity and should have the capability of process optimization to improve performance.

In this portion of the capacity assessment, plant staff should compare both the membrane bioreactor design capacity and the actual throughput achieved with the plant's average flow rates to determine the capacity status of the membrane bioreactor process. Plant staff should take into account any regulatory requirements for equipment redundancy, assumptions that may have been incorporated into system design, as well as the potential impact of cold water temperatures on membrane flux rate. The objective of this assessment is to determine if the capacity of the membrane bioreactor process is a factor limiting optimized performance of the wastewater treatment plant.

The capacity of membrane bioreactors may be most readily assessed by reviewing design specifications. Because many membrane bioreactors are newer systems, design specifications and process capacity may be readily accessible information. It is also important to compare the design specifications with the actual system's process throughput to compare theoretical capacity with the capacity realized under real-world operating conditions.

Membrane bioreactor capacity may also be quantified by multiplying the total surface area of the membranes by the membrane flux rate specified by the manufacturer, taking into account the impact of temperature on flux rate, as well as the recovery rate specified by the membrane manufacturer. This capacity is then compared with the plant's annual average flow rate for the determination of the adequacy of membrane bioreactor capacity.

Disinfection

Disinfection is typically the last step in the wastewater treatment plant process. Disinfection is applied in order to inactivate any microbial contaminants that pass through previous unit processes, in order to protect public health and environmental water quality. Although a variety of processes may be used for wastewater disinfection, the most common disinfection methods include chlorine disinfection and UV disinfection. Both of these processes are included in the Unit Process Performance Assessment in chapter 4. A discussion of the capacity assessment for both of these processes is included below. Consideration of capacity should include the ability of equipment to provide adequate disinfection, as well as the capacity of the equipment to remove any residual disinfectant present in the water, such as for the chlorination and dechlorination processes. Since UV does not produce a disinfectant residual, a similar type of residual destruction process is not required for UV systems. Plants using methods of disinfection, such as ozone or combined disinfectants, are encouraged to assess process capacity by evaluating the design capacity and functional capacity of treatment and associated chemical feed equipment, as appropriate.

Chlorine Disinfection

Chlorine disinfection is most commonly achieved through the addition of chlorine gas, purchased bulk sodium or calcium hypochlorite solution, or hypochlorite generated on site through the use of hypochlorite generation equipment. Disinfection equipment must be appropriately sized to provide adequate disinfection to meet the plant's regulatory requirements for microbial levels in the wastewater treatment plant effluent. The actual dose of chlorine that is applied to the water is dependent on the concentration and minimum contact time required, the average and peak instantaneous plant flow, and the quality of the water to which the disinfectant is applied. Organic compounds, metals, and other chemicals present in the water have the potential to exhibit a chlorine demand, so a higher quality treated water will typically require a lower chlorine dose to achieve the same target chlorine concentration as a water that has not been treated to as high a standard.

Wastewater treatment plants must comply with local regulatory requirements for the chlorination and dechlorination processes. The program recommends that plants perform the disinfection capacity assessment with regards to these requirements. If these are not available, the following general guidelines, presented in Table 3-5, are provided by Ten States Standards for recommended chlorine dosages based on water quality.

Water treated to a lesser degree, such as primary effluent, may require an even higher disinfectant dose for it to achieve adequate disinfection, due to its higher levels of disinfectant demand. The residence time in the disinfectant contact chamber must also be adequate to allow for adequate contact time, which generally is recommended at 20 to 45 minutes at average flow and 15 minutes at peak flow (Lin 2001).

The capacity of the chlorination equipment can be assessed by determining its adequacy in providing the required chlorine dose at peak flow. For example, if an activated sludge plant effluent requires a chlorine dose of 8 mg/L to treat its average flow of 10 MGD, calculate the pounds of chlorine required as follows:

$$\text{LBS/Day Required} = 8.34 \times \text{Dose Required (mg/L)} \times \text{Flow (MGD)}$$

$$\text{LBS/Day Required} = 8.34 \times 8 \text{ mg/L} \times 10 \text{ MGD}$$

In this example, the plant requires approximately 667 pounds of chlorine per day to meet its disinfection requirements.

Table 3-5. Recommended chlorine dosage for disinfection, by treatment type

Type of treatment	Recommended chlorine dosage (Ten States Standards)
Trickling filter plant effluent	10 mg/L
Activated sludge plant effluent	8 mg/L
Tertiary filtration effluent	6 mg/L
Nitrified effluent	6 mg/L

Source: Wastewater Committee of the Great Lakes–Upper Mississippi River Board of State and Provincial Public Health and Environmental Managers. 2014.

Plant staff should ensure that adequate feed capacity exists to deliver the required amount of chlorine to the effluent water, taking care to account for the concentrations of bulk purchased or on-site generated hypochlorite solutions. Staff should also ensure that basin size allows for adequate chlorine contact time prior to dechlorination by comparing the disinfectant basin size with the plant's flow rate. For example, if two 100,000-gallon basins are treating the plant's average flow of 10 MGD, a 15-minute contact time is required. To determine the basin size adequacy:

Basin Volume (MG) / Flow (MGD) = Contact Time (days)

0.2 MG / 10 MGD = Contact Time (days)

0.02 days = Contact Time

In this example, 0.02 days equates to 28 minutes, so the basin contact time is adequate.

Maintenance items that can be quickly and easily remedied should be addressed if identified during the self-assessment process. However, only long-term maintenance items that require extensive upgrade projects need to be considered performance limiting factors for the capacity assessment.

Dechlorination

In order to meet effluent permit requirements for total residual chlorine, which are often set at low concentrations (<0.1 mg/L is not uncommon) plants must commonly dechlorinate the water, after the chlorine contact time has been achieved, through the application of a dechlorinating agent. Common dechlorinating agents include sodium thiosulfate (typically applied as a solution), sodium bisulfite (typically applied as a solution), and sulfur dioxide (typically applied as a gas). The plant must have the capacity to apply a sufficient amount of dechlorinating agent to neutralize the amount of chlorine present in the water in order to achieve the plant's discharge permit requirements.

The amount of dechlorinating agent required is dependent upon the amount of residual chlorine present in the water. Table 3-6 presents guidelines for sodium thiosulfate, sodium bisulfite, and sulfur dioxide.

Table 3-6. Theoretical concentrations of dechlorinating agents required to neutralize 1 mg/L of chlorine

Dechlorinating agent	Theoretical concentration required to neutralize 1 mg/L chlorine
Sodium thiosulfate	0.56 mg/L
Sodium bisulfite	1.46 mg/L
Sulfur dioxide	0.9 mg/L

Source: Wastewater Committee of the Great Lakes–Upper Mississippi River Board of State and Provincial Public Health and Environmental Managers. 2014.

It is important to not add excessive amounts of dechlorinating agent chemical to the water; these reducing agents can consume dissolved oxygen in the water, which can be detrimental to the receiving water quality if the dechlorinating agents are applied in excess.

To assess the capacity of dechlorination equipment, be sure that the capacity of the equipment allows for adequate chemical feed to neutralize the residual chlorine present in the water. In the following example, sodium thiosulfate is used to dechlorinate an effluent water with a residual chlorine concentration of 1 mg/L at an average flow of 10 MGD. The amount of sodium thiosulfate required is as follows:

$$1 \text{ mg/L Chlorine} \times 0.56 \text{ mg/L Sodium Thiosulfate per mg/L Chlorine} = 0.56 \text{ mg/L Sodium Thiosulfate Required}$$

$$\text{LBS/Day Required} = 8.34 \times \text{Dose Required (mg/L)} \times \text{Flow (MGD)}$$

$$\text{LBS/Day Required} = 8.34 \times 0.56 \text{ mg/L} \times 10 \text{ MGD}$$

$$\text{LBS/Day Required} = 46.7$$

Provided that the dechlorination equipment is adequately sized to provide 46.7 pounds of sodium thiosulfate per day, the equipment capacity would be considered adequate and would not be a capacity-related performance limiting factor. Maintenance items that can be quickly and easily remedied should be addressed if identified during the self-assessment process. However, only long-term maintenance items that require extensive upgrade projects need to be considered performance limiting factors for the capacity assessment.

At many plants, instrumentation and control system programming helps to automate the chlorination and dechlorination process. Because effluent permit requirements can often be set at low concentrations, it is important that the detection limit of analytical equipment is sufficient to provide plant staff with meaningful data. A more detailed discussion of instrumentation and controls is provided in chapter 6.

UV Disinfection

Many wastewater treatment plants employ UV disinfection as an alternative to chemical disinfection to meet the plant's effluent requirements for microbial parameters. One advantage to UV disinfection is that it leaves no disinfectant residual, so there is no residual neutralization process, similar to dechlorination, which is required for UV disinfection.

Design standards for UV disinfection systems can be plant specific. The design of UV systems can be dependent on effluent water quality with respect to total suspended solids and UV-absorbing organic compounds, lamp selection and intensity, configuration, controls, and other factors. Plants assessing the capacity of a UV disinfection process should consider the following areas:

- Is the UV system adequately sized to provide adequate treatment at the plant's average flow rate, including any degrees of redundancy required by the local regulatory agency?
- Does the hydraulic design of the UV basins allow for plug flow conditions, without short circuiting, at the plant's anticipated hydraulic flow range?
- Is maintenance equipment adequate to prevent fouling and scaling, and to provide the necessary lamp intensity/dosage?
- Does an adequate control system exist to support UV system operation?

Plant staff are encouraged to review UV system design specifications to determine if the theoretical capacity of the UV system is adequate and to review current UV system operational data to determine if any operational factors may be limiting optimized performance. A more detailed self-assessment of UV system process performance is included in chapter 4.

UV system maintenance items that can be quickly and easily remedied should be addressed if identified during the self-assessment process. However, only long-term maintenance items that require extensive upgrade projects need to be considered performance limiting factors for the capacity assessment.

Additional, Advanced, and Proprietary Processes

This section presents general guidelines for assessing the capacity of processes not explicitly covered in this guide; these guidelines can be applied to processes such as anaerobic digesters or plant processes related to resource recovery. The guidelines may cover advanced, proprietary processes, or conventional processes.

General guidelines for the assessment of capacity include:

- Access the plant's design specifications to determine if the process design capacity is sufficient to meet the plant's annual average flow requirements.
  - Include ancillary equipment, such as associated pumps, aeration equipment, or sludge handling equipment in the analysis of capacity.
- If design specifications are not available, use process properties—such as basin volume, surface overflow rate, and other design variables—to determine the theoretical process capacity for comparison with the annual average flow rate.
- Assess whether unaddressed maintenance issues may be impacting the capacity of the treatment process.
- Do not overlook the potential impact of solids or BOD loading on process capacity.

Status

The status of the capacity of the wastewater treatment plant's major unit processes may be assessed by reviewing the following items:

- Review the performance potential graph developed using Excel or user capacity calculations to compare the projected rated capacity of primary treatment, secondary treatment, tertiary treatment, and disinfection processes against the plant's average flow rate.

  Is the capacity of any major unit process insufficient for meeting the plant's average flow requirements?

  - **Type 1:** Any unit process in which the capacity equals or exceeds the plant's average flow rate indicates a Type 1 process, and its physical size is adequate to achieve the plant's performance goals. Review chapter 4, Unit Process Performance Assessment, to determine if other aspects of this unit process may be limiting performance.
    - The self-assessment process is still quite applicable to Type 1 processes, as process that are adequately sized may still present significant opportunities for optimization in operational control, efficiency, and other factors. The fact that unit processes are considered Type 1 indicates that physical capacity should not be a factor limiting optimized performance.
  - **Type 2:** Any unit process in which the capacity is between 90 and 100 percent of the plant's average flow rate indicates a Type 2 process, and its physical size may be marginal in achieving the plant's performance goals. Review the other chapters of this guide to identify other aspects of the plant's design and operation that may be improved to limit the impact of the marginal unit processes.
    - If a unit process is determined to be Type 2, opportunities may exist to increase capacity through operational and administrative measures, such as improved process controls, solids handling, and/or reducing flows.
  - **Type 3:** A Type 3 process indicates that the physical capacity of the unit process could be a potential factor limiting the capacity of the treatment plant to be optimized. If the plant has any Type 3 processes, the overall plant flow rate may need to be reduced to the rated capability of the "weakest" unit process for the plant to achieve the desired performance goals. Review the plant's average flow rate and determine if there are ways to reduce this flow or make plans to address or expand the capacity of the process that is limiting overall plant capacity.

- Any unit process in which the capacity is rated at less than 90 percent of the plant's average flow rate indicates a Type 3 process, and its physical size is probably inadequate to completely achieve the plant's performance goals. A Type 3 facility can benefit from the self-assessment process. However, major facility modifications may ultimately be required to significantly improve performance and achieve optimization goals.

Additional process-specific capacity self-assessment questions follow. Plant staff should address the appropriate questions for processes employed at the wastewater treatment plant.

Preliminary Treatment

- Does the equalization basin volume provide adequate detention time to meet the plant's treatment goals?
- Is the plant's loading compromising its capacity and ability to optimize?

Primary Treatment

- Do limitations in primary clarifier design (volume or surface overflow rate) cause poor settling that results in inadequate treatment?
- Do limitations in sludge handling and removal equipment prevent optimized primary clarifier performance?

Secondary Treatment/Sludge Handling (address all that apply)

- Do capacity limitations in aeration basin design limit optimized performance of the activated sludge process?
- Do capacity limitations in sequencing batch reactors (SBR) design limit optimized performance of the process?
- Do capacity limitations in attached growth design limit optimized performance of the process?
- Do sludge handling pumps have sufficient capacity rating to handle the total solids volume, including with the largest pump out of service?
- Is pump efficiency routinely tested to quantify performance?
- Are any maintenance issues limiting the performance of sludge handling/removal systems in use at the plant?
- Do limitations in secondary clarifier design (volume or surface overflow rate) cause poor settling that results in inadequate treatment?
- Do limitations in sludge handling and removal equipment associated with the secondary clarifiers prevent optimized primary clarifier performance?

Tertiary Treatment (address all that apply)

- Do limitations in filter capacity limit the optimized performance of the process?
- Do any limitations in backwash system design or capacity exist that limit optimized process performance?
- Is membrane bioreactor (MBR) capacity limiting optimized performance of the system?

Disinfection

- Is the capacity of the chlorine disinfection system suitable to provide adequate disinfection?
- Is the capacity of the dechlorination system suitable to provide an adequate amount of dechlorination?
- Is the capacity of the UV system sufficient to meet the wastewater treatment plant's permit requirements?

Other Processes

- Does the capacity of any other unit process applied at the plant limit optimized plant performance?

Action

If the capacity of any of the unit treatment processes included in the major unit process evaluation does not adequately support the plant achieving the performance goals, action plans should be developed and implemented to address the identified limitations. This section offers some alternatives for consideration for addressing identified capacity deficiencies. In some cases, it may be necessary for plants to consider physical capacity upgrades to address capacity limitations. However, the following list presents a variety of actions to consider prior to initiating a project of this magnitude. Special studies may be appropriate to determine the potential impacts of these changes on plant operations and effective capacity. It is also important to consider the potential impact of any unintended consequences associated with making a change to any process. The appropriate local regulatory agency should be consulted prior to making any changes to plant process protocols that may require regulatory approval.

Primary Treatment

If primary treatment or associated sludge handling capacity is identified as a performance limiting factor, consider the following actions to address the identified capacity limitations.

- Options to reduce plant flow or loading rates to a point at which primary treatment will be adequate.
- Implementation of a sewer-use ordinance or bylaw to assist in controlling loading delivered to the treatment plant.
- Addition of a chemical coagulant or coagulant aid.
- Application of automated solids removal equipment.

Secondary Treatment or Sludge Handling

If secondary treatment or associated sludge handling capacity is identified as a performance limiting factor, consider the following actions to address identified capacity limitations.

- Explore options to reduce flow or loading rates to a point at which secondary treatment will be adequate.
- Application of process optimization techniques, such as the sludge mass control technique, discussed in greater detail in chapter 4.
- Addition of a coagulant or coagulant aid to assist in secondary clarification.
- Application of automated solids removal equipment in secondary clarifiers.
- Optimize secondary process performance to help maintain the optimal solids concentration for the sludge handling pumps.
- Implementing a proactive efficiency testing and maintenance program for critical sludge handling pumps throughout the plant.

Tertiary Treatment

If tertiary treatment capacity is identified as a performance limiting factor, consider the following actions to address identified capacity limitations.

- Explore options to reduce flow or loading rates to a point at which tertiary treatment will be adequate.
- Consider improvements to upstream processes to improve the quality of the water applied to the filters.
- Explore the feasibility of modifying the filter media composition or structure to improve the filtration process.
- Consider automating the filter backwash process.
- Evaluate membrane cleaning procedures and chemicals to determine if a change may improve flux rates and throughput (for plants using membrane bioreactors).
- Pilot studies may be used to support changes in tertiary filter operations procedures.

Disinfection

If disinfection capacity is identified as a performance limiting factor, consider the following actions to address identified capacity limitations.

- Explore options to reduce flow or loading rates to a point at which disinfection treatment will be adequate.
- Consider improvements to upstream processes to improve the quality of the water for disinfection, particularly with respect to chlorine demand.
- Consider options to increase basin detention time, such as baffles, if additional disinfectant contact time is required.
- Evaluate the use of alternative/backup disinfectants or disinfection application methods (e.g., gaseous chlorine vs. hypochlorite generated on-site) if this is required.
- Consider the use of automated monitoring and control equipment to optimize control of the disinfection process.
- Implement a proactive maintenance program that helps to ensure optimal performance of the UV system.

Other Unit Processes

If the capacity of another unit treatment process not covered in this chapter, such as a proprietary treatment process, is determined to be a performance limiting factor, a path forward should be developed to address the limitation. This may include using the manufacturer's recommendations in addition to selecting appropriate actions from the guidance previously mentioned. Documentation should be maintained to explain the rationale for changes, intended outcomes, and bench, pilot, and full-scale results.

Performance Limiting Factors Summary

The purpose of this section is to summarize the evaluated major unit process capacity factors that may be limiting optimized performance. In Table 3-3, indicate whether these factors are adequate (Type 1) or inadequate (Types 2 or 3). Factors checked as inadequate will be prioritized in chapter 8, Identification and Prioritization of Performance Limiting Factors, based on its impact on plant optimization as well as its urgency. Note that if a major unit process is assessed to be Type 2, it would typically receive an "inadequate" rating and be further prioritized based on its impact on plant performance. If a major unit process is assessed to be Type 3, it would also be checked as "inadequate" but may receive a higher prioritization ranking based on its impact on plant performance.

Table 3-7. Major unit process capacity assessment

Unit process	Questions for assessing adequacy of system capacity	Response (select one per category) Optimized and documented (or Type 1)	Partially optimized (or Type 2)	Not optimized (or Type 3)
General	Is the capacity of any major unit process insufficient for meeting the plant's average flow requirements?			
Preliminary treatment	Does the equalization basin volume provide adequate detention time to meet the plant's treatment goals?			
	Is the plant's loading compromising its capacity and ability to optimize?			
Primary treatment	Do limitations in primary clarifier design (volume or surface overflow rate) cause poor settling that results in inadequate treatment?			
	Do limitations in sludge handling and removal equipment prevent optimized primary clarifier performance?			
Secondary treatment/ sludge handling (address all that apply)	Do capacity limitations in aeration basin design limit optimized performance of the activated sludge process?			
	Do capacity limitations in SBR design limit optimized performance of the process?			
	Do capacity limitations in attached growth design limit optimized performance of the process?			
	Do sludge handling pumps have sufficient capacity rating to handle the total solids volume, including with the largest pump out of service?			
	Is pump efficiency routinely tested to quantify performance?			
	Are there any maintenance issues limiting the performance of sludge handling/removal systems in use at the plant?			
	Do limitations in secondary clarifier design (volume or surface overflow rate) cause poor settling that results in inadequate treatment?			
	Do limitations in sludge handling and removal equipment associated with the secondary clarifiers prevent optimized clarifier performance?			

Continued

Unit process	Questions for assessing adequacy of system capacity	Response (select one per category)		
		Optimized and documented (or Type 1)	Partially optimized (or Type 2)	Not optimized (or Type 3)
Tertiary treatment (address all that apply)	Do limitations in filter capacity limit the optimized performance of the process?			
	Do any limitations in backwash system design or capacity exist that limit optimized process performance?			
	Is membrane bioreactor (MBR) capacity limiting optimized performance of the system?			
Disinfection	Is the capacity of the chlorine disinfection system suitable to provide adequate disinfection?			
	Is the capacity of the dechlorination system suitable to provide an adequate amount of dechlorination?			
	Is the capacity of the UV system sufficient to meet the wastewater treatment plant's permit requirements?			
Other processes	Does the capacity of any other unit process applied at the plant limit optimized plant performance?			

*Utilities that want to evaluate the capacity of additional unit processes are encouraged to discuss these processes in this section of the self-assessment completion report. However, this is not a requirement for the Partnership for Safe Water Phase III self-assessment completion report submission.

References

Lin, Shar Dar. 2001. *Water and Wastewater Calculations Manual.* New York, NY: McGraw-Hill Publishing Company.

Tchobanoglous, George, H. David Stensel, Ryujiro Tsuchihashi, and Franklin Burton. 2013. *Wastewater Engineering: Treatment and Resource Recovery.* New York, NY: Metcalf & Eddy, Inc., McGraw-Hill.

Wastewater Committee of the Great Lakes–Upper Mississippi River Board of State and Provincial Public Health and Environmental Managers. 2014. *Recommended Standards for Wastewater Facilities.* Albany, NY: Health Research, Inc.

CHAPTER 4

UNIT PROCESS PERFORMANCE ASSESSMENT

The focus of this chapter is to assess performance limiting factors that are associated with controlling the unit processes of a wastewater treatment plant. The data gathered from plant drawings, specifications, and major unit processes evaluations, including the capacity assessment, provides the basic information needed to assess capacity-related performance limiting factors. This chapter is divided into two main sections: the liquid treatment process and the solids treatment process. Liquid treatment generally refers to the processing of incoming raw wastewater along with the subsequent treatment steps to produce a final effluent that meets the discharge permit requirements as defined by the local regulatory agency. Solids treatment generally refers to the handling and disposal of solid waste material generated as a result of the liquid treatment process (e.g., waste screening material, waste activated sludge). Although handling both liquid and solids is important in wastewater treatment, an emphasis is placed on liquid treatment processes because they have the most direct impact on effluent quality. The two main sections of this chapter are divided into subsections, which address the numerous individual unit processes that comprise the entire process treatment train at a typical treatment plant.

Although a variety of treatment technologies are included in this chapter, utilities completing the self-assessment only need to select and evaluate the processes relevant to their specific plant's treatment train. The individual unit processes are arranged in the order in which they would most likely be configured in a typical treatment train.

A major growing concern for wastewater treatment plants (WWTPs) is energy efficiency. While the primary objective of any treatment plant is to protect the downstream receiving environment by removing harmful contaminants, plant staff can take steps to achieve these goals in the most energy efficient manner possible. Energy use for a wastewater treatment facility can be divided into two major categories as follows. The first is process-related energy, which is the energy required for the explicit purpose of treating and handling the wastewater as it moves through the process. The second is nonprocess energy, which is the energy required for supporting systems (such as building heating and lighting) that does not directly contribute to the processing of wastewater. In many cases, the nonprocess energy represents a base load of energy that is generally required regardless of the volume of wastewater being treated (e.g., building lighting requirements do not fluctuate in response to changes in wastewater flow). Conversely, process-related energy often fluctuates significantly in response to changes in wastewater volume (e.g., a treatment plant influent pump station will run faster, longer, and/or with more duty pumps during peak flow events).

The efficiency of the process-related energy use is heavily dependent upon the operation and control of the unit process equipment. It is, therefore, realistic to expect that optimizing the unit processes to achieve the treatment objectives will directly benefit energy efficiency. For this reason, opportunities to improve process-related energy efficiency are discussed throughout this chapter and are interwoven with the discussion on process optimization. Conversely, nonprocess energy usage at a treatment plant would generally be similar to a typical commercial or light industrial space. For this reason, optimization of the nonprocess energy can be addressed through established best practices and guidelines for completing an energy audit, which is covered in more detail in chapter 5. Resource recovery may be a performance goal for some wastewater treatment facilities as well. These facilities are encouraged to set performance goals beyond those proposed by the Partnership for Clean Water, which may incorporate specific objectives related to resource recovery. The self-assessment process represents a baseline for facility assessment. Treatment plant staff are encouraged to use the process as a starting point for further discussion, and the self-assessment completion report provides a mechanism for capturing the outcomes of such discussions.

LIQUID TREATMENT PROCESSES

Source Water/Influent Control

Source water or influent, as it applies to wastewater treatment, refers to the substances that are deposited into the sewer collection system by the system users. Controlling these substances can be accomplished through the implementation of a sewer use agreement or bylaw. In some cases, this may require users to implement on-site pretreatment systems in order to deposit wastewater into the collection system that meets the sewer usage parameters. While these substances and pretreatment systems are outside the boundary of the central treatment plant, operations staff must be knowledgeable about them as they have a direct, and sometimes profound, impact on the treatment plant. This section will cover substances of concern in the influent, some control measures that can be implemented, and monitoring strategies to ensure the sewer use guidelines are being followed.

Nitrification Inhibiting Substances

Nitrification, as discussed further in this guide, is the process of converting ammonia in the raw wastewater to nitrate. The need for this process is becoming increasingly common at treatment facilities due to the detrimental impacts that excess ammonia can have on the environment and aquatic life. Subsequent sections in this guide discuss the challenges of maintaining a nitrification process with a focus on the treatment plant itself, which is where the operator has the most impact on the performance. However, the operator should be aware of the potential for substances in the influent that can inhibit the nitrification process at the plant. This is often overlooked when trying to troubleshoot a plant that is struggling with nitrification.

Total Sludge Mass Control Concept

Managing suspended growth systems requires utility staff to implement and maintain a stable level of process control. Stable process control is especially important when utility staff choose to, or are required to, achieve a higher level of treatment, such as nitrification. The aeration basin is the dynamo where treatment is achieved and sustained. To achieve stable nitrification, the activated sludge barrier must be scientifically managed. An adequate quantity of activated sludge is required in the aeration basin at all times to achieve the desired level of nitrification. Establishing process control conditions in the secondary unit processes that result in stable nitrification is a complex balance between carrying an adequate quantity of activated sludge to achieve the desired level of nitrification, while at the same time minimizing the mass of activated sludge in the secondary clarifier. The approach is called the total sludge mass control concept.

Figure 4-1 illustrates the total sludge mass control concept. Implementing the total sludge mass control concept represents a more comprehensive and scientific approach to managing the activated sludge barrier. The approach involves measuring the mass of activated sludge in the aeration tank and the secondary clarifier at the same time. The sum of the aeration tank mass and the secondary clarifier equals the total mass of activated sludge in the secondary unit processes. Using the sludge retention time (SRT) equation in Figure 4-1, the mass of activated sludge to waste is calculated on a daily basis.

Heavy Metals

Heavy metals in the influent is a concern for plant operators, as the treatment plant is usually not equipped to treat this type of wastewater. Heavy metals are often toxic to humans and the environment, and they can have significant detrimental impacts on the treatment plant, such as biomass toxicity and nitrification inhibition. Heavy metals are usually by-products of industrial wastewater that discharge to the collection system. If heavy metals are detected in the wastewater, the operations team should make efforts to locate the source of the discharge with an initial focus on industries. It is generally recommended that industries pretreat their wastewater for heavy metals removal prior to discharge. This is commonly accomplished with chemical precipitation processes, typically installed as a pretreatment unit at the industrial site, although other technologies are available.

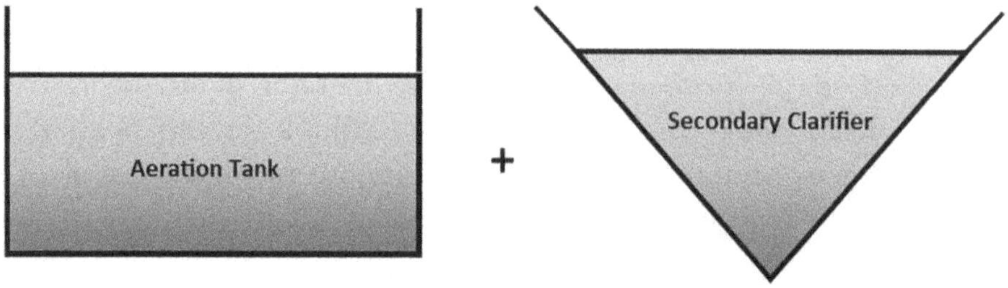

Total Sludge Mass (kg) = Aeration Sludge Mass (kg) + Clarifier Sludge Mass (kg)

$$\text{SRT (days)} = \frac{\text{Total Sludge Mass (kg) (Aeration + Clarifier)}}{\text{Sludge Mass Wasted (kg/day)}}$$

(Source: Gerard Wheeler, Global Facilitation, Inc.)

Figure 4-1. Total sludge mass control concept

Case Study: Identifying Nitrification Inhibition through Implementing the Total Sludge Mass Control Concept

City operations staff implemented the total sludge mass control concept at their WWTP in the spring of 2012. The objective of implementing a higher and more scientific level of process control was to establish process control conditions that were conducive to consistent nitrification. The program was implemented in March 2012. A period of four months passed before the facility achieved complete nitrification. Staff identified and removed multiple operations and administrative limitations between March and June 2012. The results of these activities are depicted in Figure 4-2. Starting in July, the final effluent ammonia was consistently low, with little variability, and consistent nitrification was maintained using the new process control program from July to September 2012. With consistent nitrification in place and all on-site limitations to stable process control removed, city staff were anticipating achieving stable nitrification going forward.

Starting in September 2012, the rate of nitrification unexpectedly became erratic and the final effluent ammonia concentration increased as depicted in Figure 4-3. Operating staff were surprised when the final effluent ammonia concentration increased, because the same level of process control that achieved complete nitrification from July to August 2012 was still in place in September. In response to the change, city staff began monitoring all industrial dischargers more closely to determine if a change in influent quality may have impacted the ability to maintain control of the process. In October 2013, more than one year after the final effluent ammonia concentration increased, the wastewater being discharged from an industry was confirmed to contain substances that were inhibitory to nitrification. The discovery explained the abrupt increase in final effluent ammonia that started in September 2012 and continued until October 2013. What puzzled city staff was, although the discharge of inhibitory substances from the industry stopped in October 2013, the rate of nitrification at the WWTP did not improve. This prompted city staff to investigate additional industrial facility dischargers. In January 2015, a second industry was confirmed to be discharging wastewater that was also inhibitory to nitrification. This was followed by a complete overhaul of the city's sewer-use bylaw, which included pretreatment for key industries. The new sewer-use bylaw was implemented in May 2015.

Since May 2015, the new sewer-use bylaw prevents any industry from discharging substances that are inhibitory to nitrification. Figure 4-4 includes the final effluent ammonia concentration for the period of January 2015 to May 2016. The total sludge mass control program is still applied at the plant every day. A review of the trend chart confirms that, because the new bylaw was implemented in May 2015, the frequency of elevated spikes in ammonia concentration has diminished significantly. Since June 2015, nitrification has become more stable with typical effluent ammonia concentrations in the 2 to 4 mg/L range.

A feature of the total sludge mass control program is that "cause and effect" relationships are discernable. The trend charts confirm that, since June 2015, the occasions when the effluent ammonia concentration is higher are linked with specific events at the WWTP. The arrows in Figure 4-4 depict these events and their identified causes. Plant staff members regularly discuss process performance deviations so that their cause may be identified and addressed. While no one at the city was pleased to see the periodic increase in effluent ammonia concentration, it was encouraging to know the cause of the increases were either supernatant from the digesters or low dissolved oxygen (DO) caused by an electrical failure at the WWTP, rather than a nitrification inhibition event.

The expectation for the remainder of 2016, as the new sewer-use program is fully implemented, is that the rate of nitrification will improve and the final effluent ammonia concentration will continue to remain low using the total sludge mass control concept.

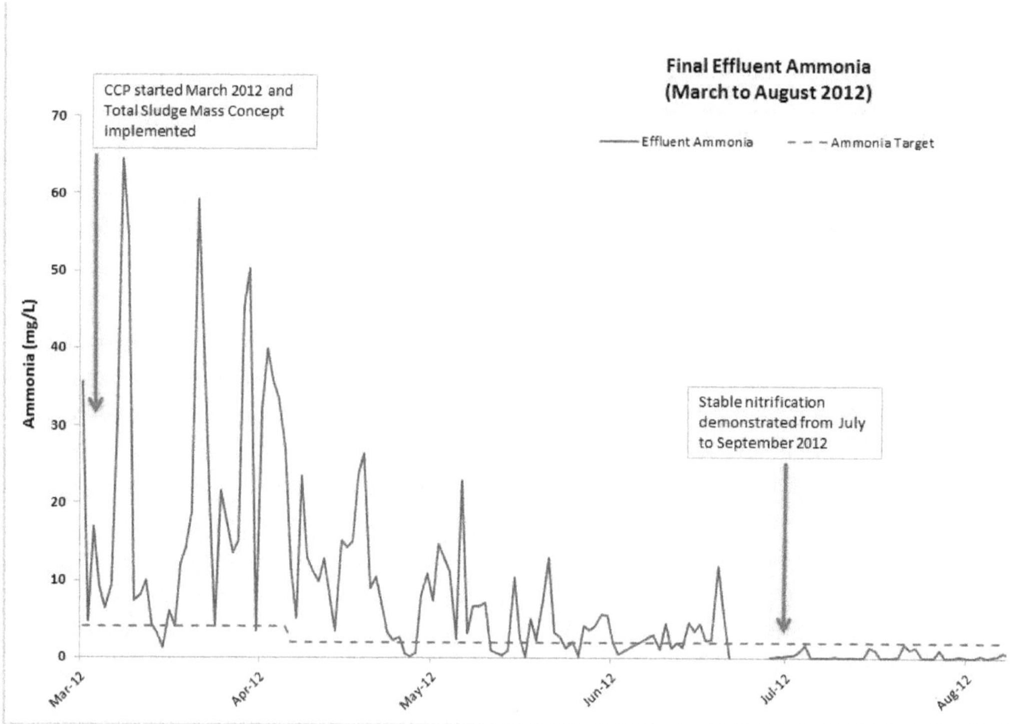

(Source: Gerard Wheeler, Global Facilitation, Inc.)

Figure 4-2. Final effluent ammonia concentration, March to August 2012

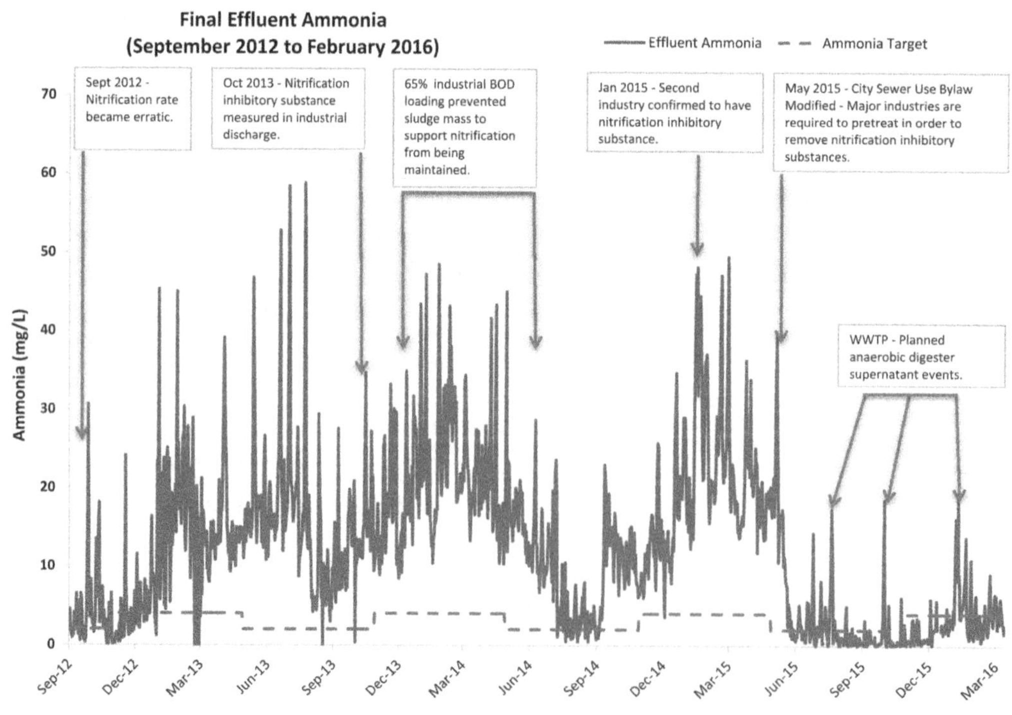

(Source: Gerard Wheeler, Global Facilitation, Inc.)

Figure 4-3. Final effluent ammonia concentration, September 2012 to March 2016

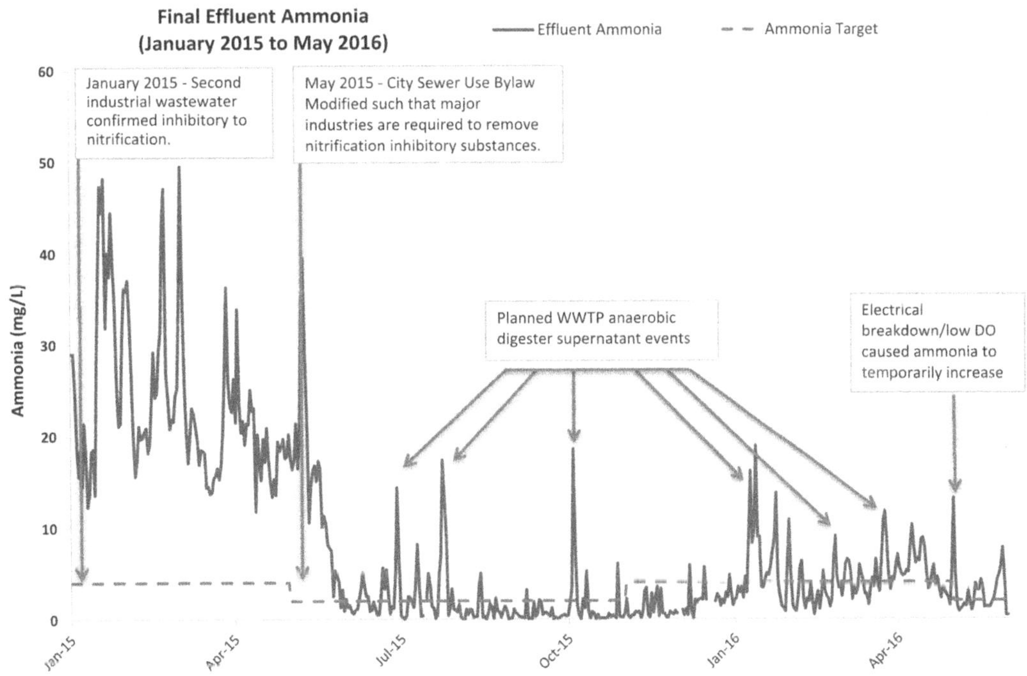

(Source: Gerard Wheeler, Global Facilitation, Inc.)

Figure 4-4. Final effluent ammonia concentration, January 2015 to May 2016

Fat, Oil, and Grease

The presence of fat, oil, and grease (FOG) in the influent wastewater can create problems at the treatment plant, such as fouling of equipment, odor issues, and unsightly conditions. Some plants may be equipped to remove FOG as part of the preliminary treatment process (described further in this chapter). A source water control program is recommended to address local industries that may require FOG removal prior to discharge. Industries and restaurants should be required to install on-site traps to allow collection and disposal of this material prior to discharge. Even at plants with FOG traps, implementing source water control will lead to improved process efficiencies at the WWTP.

Status

Effective source water/influent control is achieved through diligent monitoring and through the application of appropriate sewer-use bylaws and regulations. As was demonstrated in the nitrification inhibition case study, the plant was able to demonstrate a high level of process control due diligence through daily application of optimization concepts and accurate record-keeping. The data confirmed that the plant's loss of nitrification was a result of outside influences to the quality of the influent stream. Without a more scientific and higher level of process control, it would have been difficult to rule out the treatment process as the cause of the problem. Diligently applying the process control techniques included in this guide will enable the operations team to use data to more effectively detect harmful contaminants in the source water/influent.

The following questions are designed to help the operator better understand the status of source water/influent quality, its potential impact on plant operations, and staff preparedness for source water challenges. Operators are encouraged to develop and consider other questions that may be specific to their utility.

- Have potential risks to source water/influent quality been identified?
- Is source water/influent routinely sampled and monitored for changes in water quality?
- Has a source water control program been implemented in the community in the form of a sewer-use bylaw or similar regulation?
- Are industries that discharge potentially harmful or toxic substances required to pretreat their wastewater prior to discharge?
- Are industries required to submit regular testing and/or maintenance records to verify and demonstrate the performance of any on-site pretreatment systems?
- Is a protocol in place to respond to source water/influent emergencies?

Action

If areas of the source water/influent control program are considered partially optimized or not optimized, the following steps are recommended for implementing an action plan. In addition to this, the treatment plant should take all steps recommended in the applicable sections of this guide to optimize each and every unit process. Diligent application of process control, optimization concepts, and record-keeping, in the form of data collection and trend charts, will help confirm outside influences as the cause of issues at the plant.

- If potential risks to source water/influent quality have not been fully discussed and identified, work with utility staff and local dischargers to ensure that potential risks to water quality, that could have the potential to impact plant performance, have been defined.
- Consider implementing a source water/influent monitoring program for routine parameters as well as specific contaminants that may be of particular concern.
- If not already in place, creation and implementation of a sewer-use bylaw or similar regulation should be considered for all users of the collection system. The sewer-use bylaw should specify maximum allowable concentrations of all substances of concern.

- Industries that discharge to the collection system and exceed a defined loading limit should be required to pretreat their wastewater to remove any substances exceeding the allowable limits in the sewer-use bylaw.
  - Industries that require pretreatment should also be required to maintain operation and maintenance records to confirm the reported performance of the pretreatment plant. The treatment plant operations team or public works department staff should regularly review records from industries that are operating on-site pretreatment systems.
- If a protocol or standard operating procedures (SOPs) are not in place to address source water/influent emergencies, SOPs should be developed to address this scenario. All plant staff should be trained on the plant SOPs, which should be reviewed and updated on a regular basis to ensure that they remain current and relevant.

Performance Limiting Factors Summary

Table 4-1 summarizes factors related to source water/influent that may be limiting optimized performance of the plant. Check whether these factors are Optimized and Documented, Partially Optimized, or Not Optimized. Factors identified as Partially Optimized or Not Optimized will be prioritized in chapter 8, Identification and Prioritization of Performance Limiting Factors/Action Plan Development, and an action plan can be developed and implemented that allows for optimization of these parameters.

Preliminary Wastewater Treatment

Understanding

Preliminary treatment is typically the first step in the wastewater treatment facility for the removal of waste material. The preliminary treatment processes described in this section rely on mechanical means to separate material from the wastewater that would be detrimental to the downstream primary and secondary treatment processes. This is considered the first barrier in the multibarrier concept for wastewater treatment. Optimization of these units will have a direct positive impact on the downstream treatment processes. Typically, the preliminary treatment equipment would include some or all of the following processes.

Coarse Screening

As described in the source water/influent section of this chapter, the treatment plant is often at the mercy of the collection system and its users. A wide variety of substances, materials, and even larger debris items make their way into the collection system and ultimately end up at the treatment plant. At many facilities, coarse screening is often one of the first unit processes in the treatment train. Coarse screening serves as a physical barrier to larger debris that may be present

Table 4-1. Performance limiting factor summary: source water/influent

Self-assessment category	Questions for gauging optimization status	Optimization status		
		Optimized and documented	Partially optimized	Not optimized
Source water/ influent	Have potential risks to source water/ influent quality been identified?			
	Is source water/influent routinely sampled and monitored for changes in water quality?			
	Has a source water control program been implemented in the community in the form of a sewer-use bylaw or similar regulation?			
	Are industries that discharge potentially harmful or toxic substances required to pretreat their water prior to discharge?			
	Are industries required to submit regular testing and/or maintenance records to verify and demonstrate the performance of any on-site pretreatment systems?			
	Is a protocol in place to respond to source water/influent emergencies?			

in the raw wastewater. Coarse screening is often defined as the removal of debris larger than 6 mm in size.

Coarse screens come in a variety of styles and configurations but the ultimate goal is the same: the removal of large debris that could potentially clog or interfere with the operation of downstream pumps, aeration systems, or other critical process equipment. Coarse screens typically remove debris such as rags, cleaning pads, and other larger solid items. It is common to find coarse screens installed in a raw wastewater channel although some units are installed as enclosed tank units with inlet and outlet piped connections. The screen itself may be a bar screen or some type of perforated plate usually manufactured from carbon or stainless steel. Screens may be stationary, requiring manual cleaning or raking to remove trapped debris. Or, screens may be supplied with some type of automatic cleaning mechanism, including a motorized rake, brushes, and/or spray water. In many plants, the screenings are collected by an auger mechanism and dewatered before being deposited into some type of collection bag or bin for ultimate disposal. The filtrate from the dewatering mechanism is usually returned to the main plant flow.

Grit Removal

Grit is a common problem for many wastewater facilities and is often a result of inflow into the collection system at or near manholes or catch basins. In general, grit refers to inert solid material such as sand and gravel. Many older collection systems are prone to significant volumes of inflow and many plants also are required to handle stormwater runoff as a result of combined sewers. This can lead to significant quantities of grit material in the raw wastewater. This material is of particular concern to mechanical equipment in the treatment system, such as pumps that will wear out prematurely due to excessive abrasion resulting from grit in the wastewater. While grit generally does not pose a serious threat to the environment, it can be devastating for mechanical equipment in the plant and can heavily impact plant maintenance requirements. Therefore, it is desirable to remove grit as early as possible in the treatment train. Often, grit removal is accomplished downstream of the coarse screening to avoid the accumulation of large debris items in the grit removal system.

As indicated, grit generally consists of inert material, such as sand and gravel, that has a much higher density than water and settles readily. The two main objectives of a grit removal system are to remove the grit material and ensure that the organic solids remain in suspension so they are carried through to the downstream biological treatment process (as much as possible). Three types of grit systems are typically in use.

- Older grit systems typically consist of horizontal channels that rely on reducing the linear velocity to a suitable level (i.e., <1.0 ft/s) and providing sufficient time to allow the grit particles to settle out of the wastewater flow. If designed properly, these types of systems can be effective at removing grit, but if too much time is provided in the grit chamber during low flow periods, the system is prone to excess accumulation of organic material.
- Aerated grit chambers in which air is introduced on one side of a rectangular tank are common in the industry. The airflow is perpendicular to the direction of wastewater flow and creates a rolling action within the wastewater. If the proper airflow rate can be maintained, the heavier grit particles will settle out while the lighter organic particles will stay in suspension.
- Vortex-style grit chambers are a third type of removal system prevalent in wastewater facilities. Flow enters a circular grit tank tangential to its perimeter and flows in a circular path around the outer wall of the tank. The flow path induces a vortex within the chamber that helps to promote separation of the heavier grit particles from the lighter organic particles. The inlet is located below the outlet to prevent grit from being carried through the system. The relatively high velocities and the action of the induced vortex help to keep the organics in suspension so they can pass freely to the outlet of the grit chamber. In some cases, depending on the hydraulic characteristics and the variability in the influent flow rates, vortex-style grit chambers are outfitted with motorized paddles that rotate within the chamber to ensure the required velocity is maintained.

Regardless of the type of grit chamber used, it is common for the system to include a grit slurry handling and dewatering system. In the case of horizontal or aerated grit chambers, the use of a mechanical grit removal system, such as a motorized rake or inclined auger system, can be used to extract and dewater the grit prior to final disposal. Vortex grit chambers often use airlift pumps or motorized grit pumps to draw the girt slurry from the bottom of the chamber and direct it to the grit dewatering unit. Often, an inclined auger-style screen is used to dewater the grit. Most grit extraction and dewatering systems also include the capability to further separate organic material and return this to the main wastewater flow. This type of system is referred to as grit classification.

Fine Screening

Fine screening is a second screening step that may be present in some facilities; it generally refers to the removal of solids and debris that are less than 6 mm in size. Fine screening for some facilities can remove material down to 1 mm or less if required. Fine screens provide a physical barrier to the solids, usually by way of a submerged wire mesh or perforated plate. Fine screens typically include a mechanical cleaning system, such as a rotating brush assembly and/or a water spray. Similar to coarse screening, the fine-screened solid material is typically removed from the system and dewatered prior to final disposal.

Fine screening, particularly at finer pore sizes, is usually only included at facilities that have a specific need to do so. As the screen opening becomes finer, there is a greater chance that organic material will be excessively removed from the system. It is undesirable to remove excess organic material prior to the secondary treatment process unless a specific process-related reason exists for doing so. A good example of where fine screening is necessary is with membrane bioreactors (MBRs). The MBR process, as described further in this guide, uses very fine pore membranes (i.e., <1 micron) as part of the secondary treatment process to remove virtually all total suspended solids (TSS) from the treated effluent. The delicate nature of the membrane material means they are susceptible to damage and fouling from solids in the wastewater. MBR systems are almost always designed with fine screening as protection for the membranes, and the pore size of the fine screens is often less than 3 mm (the actual requirements depend on the membrane manufacturer).

Grinders and Grinder Pumps

Some facilities use grinding equipment as a part of their treatment plant headworks or preliminary treatment process. Grinders are different from screening equipment in that they do not remove any debris from the raw wastewater. Material is physically cut or ground into smaller pieces. Grinding is usually included to allow larger debris to pass more easily through downstream pumping equipment. In some cases, the grinding function may be incorporated into the design of the raw wastewater pumps in the form of a grinder pump. In this case, the cutting element is incorporated into the pump to allow material to be reduced in size as it enters the pump.

Fat, Oil, and Grease Removal

The presence of fat, oil, and grease (FOG) in the raw wastewater can create issues at treatment plants, including fouling or clogging of equipment, odors, and unsightly appearance. As identified in the source water/influent control section, generators of excess FOG to the collection system, such as restaurants and industries, should be required to install equipment on-site to remove this material from their discharge. Source water/influent control, through an effective sewer-use bylaw, is the most efficient means for the plant to control this material and prevent it from causing issues. However, in some cases, even with good source water control, plants may find they still have issues with this type of material. Dedicated traps can be installed at the plant to remove FOG material from the raw wastewater as it enters the plant. Other options may include the installation of skimmers and troughs in equalization or primary clarifier tanks.

Status

The preliminary treatment equipment, regardless of the type of technology chosen, consists of mechanical processing of the raw wastewater. The design of these units is critical for successful operation. A thorough review of the design capacity, as addressed in chapter 3, is the first step in optimization. In some cases, the only course of action may be to install a larger unit to properly handle the hydraulic loading. Because this equipment is typically installed upstream of any equalization tank, it typically is sized to handle the peak flow rate.

If the equipment is determined to be adequately sized, the primary method for optimization is to implement a good inspection and maintenance program. Many of the mechanical components associated with the preliminary treatment are exposed to harsh operating conditions and are subject to significant wear and tear through normal operation. Regular inspection and replacement of worn components, in combination with following manufacturer recommended maintenance schedules, will help promote optimal performance. Typically, a regular inspection program will provide the operators with a visual impression of the performance of the various units. For example, the operator may notice a change in the quantity of screenings being collected in the waste bin that could be a sign of reduced performance. Excessive wear of some components can result in material bypassing the screen (material that otherwise would have been captured by a well-operating machine). An excessive amount of debris in downstream processes, such as an equalization tank or primary clarifier, may also be an indication of material bypassing the screen.

The following questions are designed to help the operator better understand the status of preliminary wastewater treatment processes and equipment.

Operators are encouraged to develop and consider other questions that may be specific to their utility:

- Has a regular maintenance and inspection program been developed for the facility that includes the preliminary treatment equipment? Inspection items may include the following:
  - ▲ Visual inspection of material being removed from screening or grit removal equipment.
  - ▲ Regular checking and emptying of oil and grease traps as required.
  - ▲ Regular inspection of common wear parts on all equipment.
  - ▲ Inspection and testing of any backflush or wash water lines and nozzles.
  - ▲ Regular lubrication of mechanical equipment as recommended by the manufacturer.
  - ▲ Manual removal of any abnormally large debris that may cause damage to equipment.
- Does the plant maintain an inventory of spare parts to minimize equipment downtime?
- If the equipment is designed in a duty/standby configuration, is there a policy to ensure equal runtimes on each unit?
- Has the plant optimized the source water control program to minimize fat, oil, and grease material from being discharged into the collection system? This will improve the performance and efficiency of all downstream equipment being operated at the treatment plant.

Action

If areas of the preliminary treatment operation are considered Partially Optimized or Not Optimized, the following steps are recommended for consideration in developing and implementing an action plan.

- The plant should implement a regular inspection and maintenance program for all equipment. Usually, little or no sampling and testing are completed to monitor the performance of the system, so an effective inspection and maintenance program is the only method the operator can use to gauge equipment performance.
- Regularly inspecting screening and grit waste material quantities can help identify reduced performance. Larger than usual material downstream of a grinder or grinder pump may indicate excessive wear on the cutter(s).

- Operators should diligently follow regular maintenance schedules recommended by the manufacturers and replace worn mechanical components as soon as possible.
- The plant should maintain an inventory of spare parts as recommended by the manufacturer and ensure the inventory is replenished as needed.
- It is recommended that the plant control system monitor and record equipment runtimes. In some plants this may be helpful in tracking maintenance requirements for the equipment. Also, in plants where equipment operates in a duty/standby arrangement, the control system should automatically alternate the duty unit on a regular cycle to promote even runtime and wear on all units.
- It is recommended that the plant implement a source water/influent control program, such as a sewer-use bylaw as recommended in this guide. Any efforts to reduce fat, oil, and grease material at the source will lead to improved efficiency at the plant.

Performance Limiting Factors Summary

Table 4-2 summarizes factors related to preliminary wastewater treatment that may be limiting optimized performance of the plant. Check whether these factors are Optimized and Documented, Partially Optimized, or Not Optimized. Factors identified as Partially Optimized or Not Optimized will be prioritized in chapter 8, Identification and Prioritization of Performance Limiting Factors/Action Plan Development, where an action plan can be developed and implemented that allows for optimization of these parameters.

Primary Wastewater Treatment

Primary treatment generally refers to processes downstream of any screening and grit removal processes but prior to any biological secondary treatment. Common primary treatment unit processes include equalization and primary clarification (or sedimentation). The main purpose of these processes is to control the hydraulic, organic, and solids loading on the downstream secondary treatment processes. Equalization tanks/basins provide a more consistent flow and loading to the downstream secondary treatment process through storage volume and attenuation. Equalization does not provide any removal or reduction. By contrast, primary clarification assists the downstream secondary process by physical removal of contaminants, primarily BOD_5 and suspended solids.

Equalization

Understanding

At many WWTPs, numerous factors—including seasonal changes, weather events, and normal diurnal fluctuations—can have a significant impact on flow

Table 4-2. Performance limiting factor summary: preliminary treatment

Self-assessment category	Questions for gauging optimization status	Optimization status		
		Optimized and documented	Partially optimized	Not optimized
Preliminary treatment	Has a regular maintenance and inspection program been developed for the facility that includes the preliminary treatment equipment?			
	Does the plant maintain an inventory of spare parts to minimize equipment downtime?			
	If the equipment is designed in a duty/standby configuration, is there a policy to ensure equal runtimes on each unit?			
	Has the plant optimized the source water control program to minimize fat, oil, and grease material from being discharged into the collection system?			

variability. In some cases, depending on the event, peak flow rates can be several times greater than the average day flow. Treatment plants, particularly those with biological processes, operate optimally when the flow and organic loading is consistent. Extreme peak flow events can result in excessive organic loading to the process, solids washout due to high overflow rates in the secondary clarifier, hydraulic overloading of pumping equipment, and potential discharge of untreated wastewater to the environment. Plant staff must properly handle these peak flow events. However, it is often impractical to design a biological process for short-term peak flows. At many plants, the solution is to include capacity for attenuation of flows in what is commonly referred to as an equalization tank/basin. The equalization tank typically is in the main liquid train immediately downstream of the headworks equipment. If properly sized, the tank provides sufficient volume to store excess wastewater during peak flow events. This allows wastewater to be pumped consistently to the downstream treatment processes even during peak flow events. When the short-term peak flow event has subsided, the stored volume of wastewater can be fed to the treatment plant under controlled conditions.

Equalization tanks are typically equipped with some form of mixing to prevent solids settling while the wastewater is being stored. In many cases, this mixing is achieved by a diffused aeration system that provides mixing as well as sufficient oxygen to prevent septic conditions.

Status

In general, the equalization tank is a relatively simple unit process that requires few operational controls for operators to optimize its performance. The most critical control that the operator must manage is the tank level during normal or low flow periods. During these periods, the equalization tank level should be at a minimal wastewater volume, with the goal of keeping the level as low as possible. By maintaining the level as low as possible during normal flow periods, the operator is ensuring the maximum possible storage volume is available to provide a hydraulic buffer during periods of peak flow. The operator should also strive to empty the tank to the lowest level as soon as possible after a peak flow event, while at the same time maintaining a controlled influent flow rate to the biological process. Flow from the equalization tank to the downstream process can be accomplished by gravity or pumping depending on the hydraulic conditions at each plant. If pumps are used, then the operation should be optimized to allow for consistent flow to the biological process.

The operator also has control over the mixing and aeration systems and should monitor these on a regular basis. If persistent solids deposition occurs on the bottom of the tank, it could be a sign of inadequate mixing. Significant solids deposition can reduce the effective volume of the tank and reduce its capacity to attenuate peak flows. Aeration is provided for mixing only and, in many cases, there is no DO feedback to control the aeration blowers. The operator should check for mixing patterns, bubble dispersion on the tank surface, and odors that may indicate septic conditions to determine if the aeration level is sufficient.

The following questions are designed to help the operator better understand the status of an equalization tank/basin. Operators are encouraged to develop and consider other questions that may be specific to their utility:

- Has a low operating level target been defined for the equalization tank?
    - The low operating level should be as low as possible in the tank to provide maximum storage volume for attenuation. Some minimal level will need to be maintained so as not to cause any issues for the aeration or mixing system(s).
- Does the tank effluent flow method (pump versus gravity) and control strategy allow flexibility for consistent flow to be provided to the biological system even under peak flow events? If the pumps are equipped with VFDs, then selecting an appropriate effluent flow rate is critical. Staff can do this by comparing the instantaneous inflow and outflow diurnal curves for peak flow conditions.

- Has a routine inspection program been developed? Inspections should include the following:
  - Checking for excess solids deposition. Use of a sludge judge or sludge level analyzer may suffice or it may be necessary to periodically drain the tank, if required.
  - Visually checking for sufficient aeration and/or mixing.
- Have SOPs been developed for the equalization tank, including an emergency response plan?

Process Energy Considerations

The major energy considerations for the equalization tank are the mixing system and tank effluent pumping. Mixing systems for tanks require a theoretical amount of energy based on the tank volume and geometry. If the tank volume remains relatively consistent, so does the energy requirement for mixing. This becomes a challenge for equalization tank mixing operations in which the volume is highly variable. Properly designed mixing systems for equalization tanks should be designed to vary the mixing intensity based on control feedback from online level sensors in the tank. Varying the mixing intensity can be accomplished by reducing the number of operating blowers or mechanical mixers, or by reducing their speed through the use of variable frequency drives. At very low tank levels, it may be possible to further reduce energy consumption by cycling the mixing system on and off.

The main goal of the effluent pumping system (if pumps are used as opposed to gravity flow) is to achieve a consistent flow rate to the downstream processes. This type of operation is typically beneficial from an energy standpoint as it should allow for operation at or near the pumps' best efficiency point, assuming the pumps were properly sized during the plant design. The operator should be aware of the best efficiency point (BEP) for the pumps and strive to operate as close as possible to this point to achieve maximum efficiency. If this becomes unrealistic due to improper pump sizing, it may be advantageous to replace the pumps with properly sized units.

Additional status questions associated with equalization basin process energy considerations include the following:

- Is the mixing intensity for the equalization tank controlled based on tank level?
- Do the equalization tank effluent pumps operate at or near the BEP of the pump curves on a consistent basis?

Action

If areas of the equalization tank operation are considered Partially Optimized or Not Optimized, the following steps are recommended for developing and implementing an action plan.

- Develop operating targets for the tank, including optimizing the normal operating level for maximum attenuation storage.
- If pumping is used for tank effluent flow, select a target flow rate that allows for consistent flow to the biological system while not allowing the tank to become overloaded organically and hydraulically.
- Implement a regular inspection routine for the equalization tank that includes the following items:
  - Inspect the tank mixing patterns to ensure the tank is being sufficiently mixed.
  - Check for solids deposition on the bottom of the tank, which may be an indicator of inadequate mixing.
- SOPs for the equalization tank should be developed, including an emergency response plan.

Primary Clarification

Understanding

As with all of the unit processes included in this guide, the first step in optimization is defining unit process performance goals and treatment objectives. Primary clarification is typically designed to achieve 25 to 40 percent removal of BOD_5 and 50 to 70 percent removal of total suspended solids. Removal is accomplished through physical settling by gravity as the wastewater flow passes from the inlet to the outlet of the primary clarifier tank. Note that primary clarifiers are typically of a rectangular or circular configuration. In the rectangular configuration, the flow is generally linear from one end of the tank to the other. For circular clarifiers, wastewater commonly enters through a central pipe/structure and migrates to the perimeter of the tank in a radial manner. In either the rectangular or circular configuration, the clarified wastewater typically exits the tank by flowing over an outlet weir. For rectangular tanks, one or several linear weir structures may be used at the outlet end of the tank. Circular tanks typically have one continuous weir that encircles the perimeter of the tank. In addition to removal of settled solids, floating scum or grease that is not removed in the raw sludge is typically removed by skimmers. A waste solids pumping system is typically used to remove settled material from the tanks for stabilization, dewatering, and disposal.

In general, both tank designs rely on reducing the velocity between the inlet structure and outlet weir(s) to provide sufficient time to allow suspended solids to be settled out by gravity. In theory, if the velocity of wastewater can be controlled within an acceptable range, the performance of the clarifier should always remain within the typical operating removal range. In practice, however, it is difficult to measure velocity directly within a tank. Therefore, staff typically use alternative operational parameters, which can be calculated from the influent flow rate and the geometry of the tank. Parameters such as hydraulic retention time (HRT) and overflow rate are key design parameters for primary clarifiers. HRT is defined as the average time the wastewater spends within the basin—traveling from the inlet to the outlet—and is calculated by dividing the total tank/basin volume by the influent flow rate. Overflow rate is calculated as the flow rate divided by the total liquid surface area of the tank. Operators have no control over the geometry of existing tanks or basins, so they may have little control over these parameters, in many cases. However, calculating these parameters as flow rates change provides further understanding of the conditions under which the clarifiers are performing. In plants where multiple clarifiers operate in a parallel configuration, the calculation of these parameters can prompt the operator to put additional units into operation.

Status

While the ultimate objective is to achieve the treated effluent standards, it is desirable that the performance of each unit process within the treatment plant is also optimized. This will ensure utility staff focus on optimizing the performance of each major unit process. There are typical industry performance standards, or internally developed plant-specific goals, that each unit process is expected to achieve. Achieving the performance standards for each unit process ensures the multiple barrier concept, illustrated in Figure 2-1, is being maximized.

The following questions are designed to help the operator better understand primary clarifier status. Operators are encouraged to develop and consider additional questions that may be specific to their utility.

- Have primary clarifier removal targets (high and low range) been defined?
  - The primary removal target for primary clarifiers is TSS removal.
  - A secondary target would be BOD_5 or COD which is typically a function of the TSS removal performance.
- Does the plant sampling and testing program provide the required data to fully analyze the status of the primary clarifiers?
  - Recommended sampling, testing, and data collection includes the following items:
    - Clarifier influent and effluent TSS concentration.
    - Clarifier influent and effluent BOD_5 or COD concentration.
    - Clarifier influent flow-rate data.

- Have the appropriate trend charts been developed to visualize the performance of the primary clarifiers over time?
  - The trend chart should include the test data for TSS concentration and BOD_5 or COD concentration (if available) as well as the range (high and low) for the defined performance goals.
- Has the operator collected and analyzed primary clarifier flow data?
  - Flow data for the primary clarifier along with basic geometry of the tanks can be used to calculate key operating parameters as follows:
    - Primary clarifier hydraulic retention time.
    - Primary clarifier overflow rate.
  - It is recommended that the primary clarifier flow data and key operating data, such as HRT and surface overflow rate, be plotted on the same trend chart with the performance data. Periods of declined performance may be correlated with the flow rate, hydraulic retention time, and/or overflow rate to determine if any of these operating parameters exceed the recommended design ranges.
- Does the primary clarifier have a regular solids wasting program, and does the operator monitor key settled solids wasting parameters? Key parameters for the solids wasting that should be monitored and recorded include the following:
  - Settled solids removal rate.
  - Settled solids pump cycle times and durations.
  - Total solids concentration of the settled solids.
- Has the primary clarifier(s) been inspected for any visual signs or indicators of other potential issues as follows:
  - Excessive scum or floatable accumulation.
  - Debris buildup on outlet weir.
  - Excessive accumulation of settled solids.
  - Floating sludge at or near the liquid surface.
  - Gas bubbles.
  - Strong septic odor.
- Have SOPs been developed for the primary clarification process, including an emergency response plan and maintenance procedures?

Case Study: Using Trend Charts to Track Primary Clarifier Performance

One of the first steps in defining the status of any unit treatment process is to define the performance goals. Once the goals have been defined, the critical data are collected. A review of the data enables the actual performance to be measured and compared to the stated goals. A typical performance goal for conventional primary clarifiers would be to achieve 25 to 40 percent removal of biochemical oxygen demand (BOD_5), and 50 to 70 percent removal of TSS. The use of trend charts for monitoring performance data is a recurring theme that is repeated on multiple occasions throughout this guide.

Key data required to determine the status of a primary clarifier include the influent and effluent BOD_5 and TSS concentrations. At many plants, it may not be common to test for selected parameters at intermediate points within the system. Often, it is only the plant influent and effluent data that are collected, and typically this is done to satisfy regulatory reporting requirements. However, if the goal is to optimize individual processes, such as a primary clarifier, additional data may need to be added to the regular sampling and testing program. Once the data are collected, removal percentages can be computed and plotted over time to establish a performance baseline and demonstrate long-term trends.

Figure 4-5 demonstrates the use of trend charts. This case study pertains to primary clarifier TSS removal efficiency. At this particular plant, two liquid trains are operated in parallel, identified as Plant A and Plant B, respectively. Figure 4-5 illustrates the TSS removal percentage data for both trains along with the high and low range removal targets (70 percent and 50 percent removal, respectively). From January to December, the average TSS removal efficiency was 71 percent in the Plant A primary clarifier, and 74 percent in the Plant B primary clarifier. Both primary clarifiers achieved a high level of TSS removal efficiency during this time.

Some trends may be observed in the chart displayed in Figure 4-5. From January to April 2015, the average TSS removal efficiency was 58 percent and 62 percent for Plants A and B, respectively, which was within the typical 50 to 70 percent removal range. Starting in May 2015, the average TSS removal efficiency increased significantly in both clarifiers. The removal efficiency averaged 77 percent and 79 percent, in Plants A and B, respectively, for the remainder of 2015.

The $cBOD_5$ removal data for the same plants and time period are presented in Figure 4-6. The $cBOD_5$ trend generally follows the same pattern as the TSS trend, which is to be expected for a primary clarifier. From January to April 2015, the average $cBOD_5$ removal efficiency was within the typical 25 to 40 percent range for both primary clarifiers. Starting in May

2015, the average $cBOD_5$ removal efficiency increased significantly in both Plants A and B. The removal efficiency averaged 55 to 58 percent in Plants A and B, respectively, for the remainder of 2015. Consistently achieving, and in many months exceeding, the typical TSS and BOD_5 removal criteria demonstrates that the performance of the primary clarifier barrier is optimized. Although primary clarifier performance has been shown to be optimized for these parameters, plant staff optimally would continue to collect and trend primary clarifier TSS and $cBOD_5$ data to continue to establish long-term trends, further quantify plant performance, and be well-equipped to quickly identify any future performance deviations.

The trend chart provides utility staff with a powerful tool to summarize and interpret data along with a visual indicator of the performance of the unit process. If the performance of the primary clarifier does not achieve the desired TSS or BOD_5 removal goal, further investigation would be required to determine the causes and action steps to improve the performance.

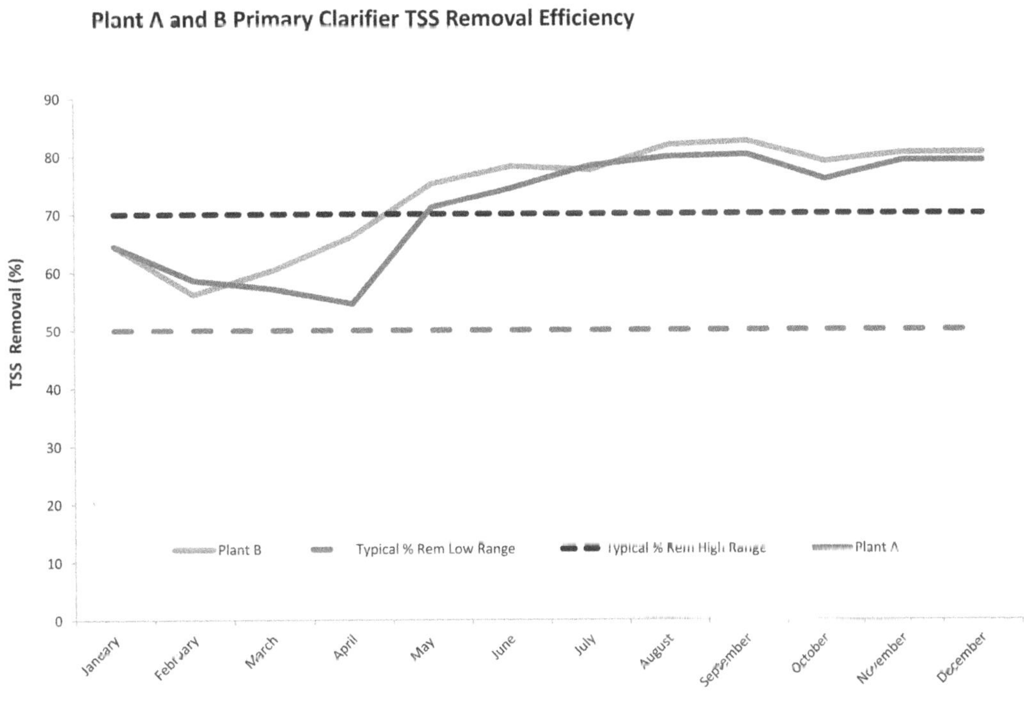

(Source: Gerard Wheeler, Global Facilitation Inc.)

Figure 4-5. Primary clarifier TSS removal efficiency

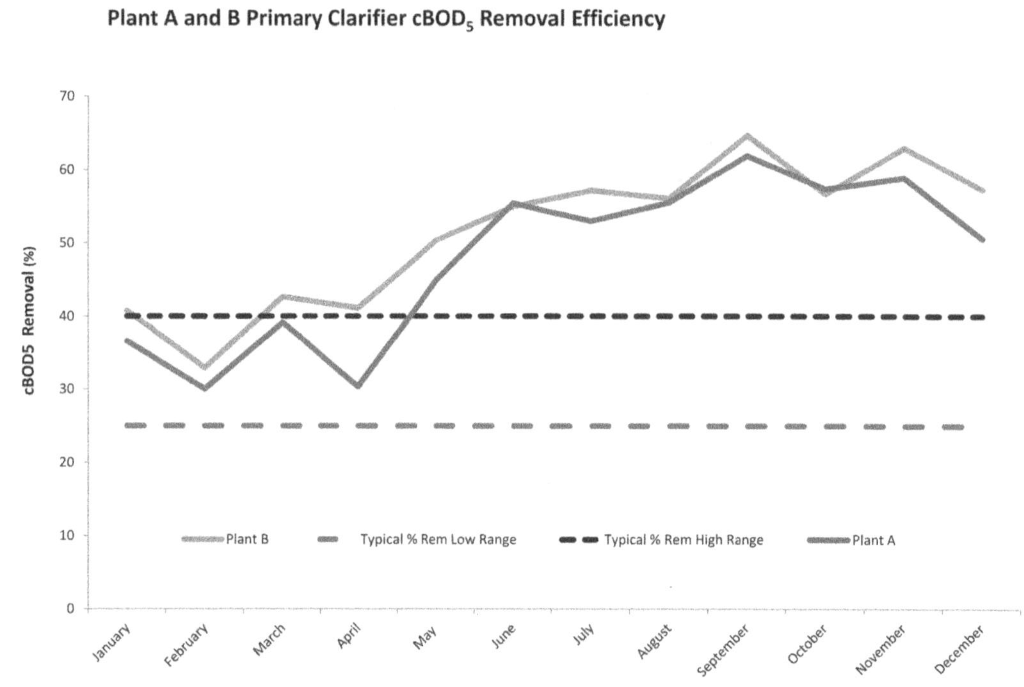

(Source: Gerard Wheeler, Global Facilitation, Inc.)

Figure 4-6. Primary clarifier $cBOD_5$ removal efficiency

Action

If areas of the primary clarifier operation are considered Partially Optimized or Not Optimized, the following steps are recommended for consideration in developing and implementing an action plan.

- Removal targets should be clearly defined for the primary clarifier unit in terms of percentage removal of TSS and/or BOD_5. The operators should consult design manuals or specifications, if available, for the existing clarifier units to determine if they were designed for a specified removal rate. If no design information is available, the following removal targets are recommended:
  - TSS: 50 to 70 percent removal.
  - BOD_5: 25 to 40 percent removal.
- Operational targets should be clearly defined for the primary clarifier in terms of hydraulic retention time. The operators should consult design manuals or specifications, if available, for the existing clarifier units to determine if they were designed for specific operating ranges.
- A sampling and testing program should be developed for the plant that includes the following information for the primary clarifier.
  - Primary clarifier influent flow rate.

  - Clarifier influent and effluent TSS and BOD_5 concentration.
  - Settled solids TSS concentration.
  - Settled solids flow rate and total daily volume.
- All pertinent data for the primary clarifier should be collected and captured on a trend chart to allow staff to visualize the unit's performance.
- A settled solids removal program should be developed for the primary clarifier. It is recommended that the settled solids be removed on a regular and consistent basis as much as possible. Staff should take care to not over-waste the solids as this can reduce the concentration of the wasted material, leading to decreased performance in the subsequent dewatering process. In general, the following guidelines are recommended for the settled solids removal system:
  - Settled solids should generally be removed from the primary clarifier a minimum of every two to four hours. If gas bubbles, floating sludge, or a strong odor are detected, it may indicate wasting that is too infrequent and is creating septic conditions within the clarifier.
  - The solids wasting should not be so frequent or lengthy that it allows the TSS concentration of the waste material to be too low. The wasting schedule should be designed to maintain a settled solids concentration in the ideal range of 1 to 2.5 percent.
- The Partnership for Clean Water recommends implementing a daily inspection routine for the primary clarifier, which should include the following items.
  - Visually inspect all weirs and remove any debris that may have accumulated.
  - Visually inspect any scum collection or skimming devices for smooth operation and take note of any excessive buildup of foam or scum.
  - An inspection log should be maintained for the primary clarifier to document any issues, actions taken, and results of any actions.
- SOPs for the primary clarifier should be developed, including an emergency response plan and maintenance procedures.
  - The operator is encouraged to take an active role, along with plant management, in developing SOPs.
  - The operator should regularly review the procedures and the emergency response plan to be better prepared in the event of an unexpected issue with the system.
  - The operations team is encouraged to continually update and improve SOPs and response plans, and help ensure that all staff are trained in these procedures.

- Well-documented SOPs will provide better continuity in the event of staff changes within the facility.

Performance Limiting Factors

Table 4-3 summarizes factors related to the primary wastewater treatment processes, including equalization and primary clarification, which may be limiting optimized performance of the plant. Check whether these factors are Optimized and Documented, Partially Optimized, or Not Optimized. Factors identified as Partially Optimized or Not Optimized will be prioritized in chapter 8, Identification and Prioritization of Performance Limiting Factors/Action Plan Development, where an action plan can be developed and implemented that allows for optimization of these parameters.

Secondary Wastewater Treatment

This section of the chapter covers common secondary wastewater treatment processes, such as CAS, attached growth processes, and secondary clarification. Each section includes a set of self-assessment questions. Plant staff need only address the self-assessment questions relevant to processes applied at their treatment facility. Where appropriate, case studies have been included to provide additional perspectives on process optimization techniques and results.

Suspended Growth (Conventional Activated Sludge)

Understanding

At many wastewater facilities, the key treatment process is based on the conventional activated sludge (CAS) process, which is a suspended growth secondary treatment process for the biological oxidation of biochemical oxygen demand (BOD_5). The CAS process uses aerobic bacteria to consume the biodegradable organic material (i.e., BOD_5) in the incoming wastewater. The process is usually contained within concrete tanks where a specific concentration of bacterial population, referred to as the biomass, is maintained. Wastewater is fed to the bacteria and the combined tank contents, commonly referred to as the mixed liquor, are aerated to provide the oxygen required to sustain the biomass. Air is usually introduced through submerged air diffusers supplied from mechanical aeration blowers. It is common for aeration systems to be designed to maintain a DO concentration in the mixed liquor of approximately 2.0 mg/L; this ensures sufficient air is being provided to the process. In the presence of sufficient air and proper temperature conditions, the CAS system may be capable of converting ammonia in the wastewater to nitrate, a process commonly referred to as nitrification. This additional treatment capability will be discussed further in the nutrient removal section that follows.

Table 4-3. Performance limiting factor summary: primary wastewater treatment

Self-assessment category	Questions for gauging optimization status	Optimization status		
		Optimized and documented	Partially optimized	Not optimized
Equalization	Has a low operating level target been defined for the equalization tank?			
	Does the tank effluent flow method (pump versus gravity) and control strategy allow flexibility for consistent flow to be provided to the biological system even under peak flow events?			
	Has a routine inspection program been developed?			
	Have SOPs been developed for the equalization tank, including an emergency response plan?			
Primary clarification	Have primary clarifier removal targets (high and low range) been defined?			
	Does the plant sampling and testing program provide the required data to fully analyze the status of the primary clarifiers?			
	Have the appropriate trend charts been developed to visualize the performance of the primary clarifiers over time?			
	Has the operator collected and analyzed primary clarifier flow data?			
	Does the primary clarifier have a regular solids wasting program, and does the operator monitor key settled solids wasting parameters?			
	Has the primary clarifier(s) been inspected for any visual signs or indicators of other potential issues?			
	Have SOPs been developed for the primary clarification process, including an emergency response plan and maintenance procedures?			
Process energy considerations – Equalization Basin	Is the mixing intensity for the equalization tank controlled based on tank level?			
	Do the equalization tank effluent pumps operate at or near the BEP of the pump curves on a consistent basis?			

The CAS system is based on the use of living organisms to break down the organic material in the wastewater. The process requires certain operating conditions to help the organisms thrive within the system. As mentioned previously, oxygen supplied through aeration is a critical operating parameter and is typically monitored by the measuring the mixed liquor DO concentration. The organisms also require a certain mass of "food" (i.e., BOD_5) in order to sustain themselves and grow the population. Typically, this is monitored through two key operational parameters identified as the OLR and the food-to-microorganism ratio (F:M). Organic loading rate (OLR) is defined as the daily mass of BOD_5 fed to the system per unit volume within the aeration tank(s). While the OLR is a useful parameter to monitor, it does not take into account the concentration of the biomass within the system. The plant F:M ratio better quantifies this as it is defined as the mass of BOD_5 per unit of biomass in the system. The biomass is normally quantified as the mass of volatile suspended solids in the aeration tank(s), and it is often referred to as the mixed liquor volatile suspended solids or MLVSS. Multiplying the MLVSS concentration by the total volume of the tank will give an estimate of the biomass in the aeration tanks (typically expressed in units of pounds or kilograms). The mass of BOD_5 being fed to the system can be calculated by multiplying the influent concentration, as measured at the inlet to the aeration tanks, by the influent flow rate (usually expressed in units of pounds or kilograms per day). Both the OLR and the F:M ratio help the operator to determine if the correct amount of biomass is being maintained within the system for the measured mass of BOD_5 being fed on a daily basis. If too much BOD_5 is being introduced and sufficient biomass inventory is not being maintained, the system is organically overloaded, which can be harmful to the organism population.

In many plants, the operator may have limited control over the concentration of BOD_5 coming into the secondary treatment process. Optimizing the performance of the primary clarifier (discussed previously) will help to control the organic loading to the secondary process; however, many operators have little-to-no control over what is actually disposed to the collection system and sent to the treatment plant. Even with proper control of the primary clarification process, significant fluctuations can occur in the flow and organic loading to the secondary treatment process. The operator needs to take steps to ensure that a sufficient inventory of biomass is available in the aeration tanks to handle these fluctuations. The primary method of controlling the biomass inventory at many facilities is through wasting of excess sludge. As the biomass population feeds on the organic material in the wastewater, it grows through synthesis of new cells. Older cells will die off and must be removed by wasting. If the growth in biomass exceeds the rate of wasting, there will be a net-positive growth in biomass generation. Over time, some cells are removed from the system, either in the effluent or through removal of WAS. At most facilities, wasting of activated sludge is based on achieving a target mixed liquor suspended solids (MLSS) concentration in the aeration tank(s). By operating only using a target MLSS concentration to control wasting, many operators unknowingly overlook the solids retention time (SRT) for the system. Operation of the secondary treatment process works best if the sludge mass control system includes both a target MLSS concentration and a target SRT.

A number of process variants based on the CAS process are available, including step-feed, extended aeration, oxidation ditches, and sequencing batch reactors (SBR). Many of these processes apply the same general biological principles as the CAS process. The primary difference would be the typical design criteria for OLR, F:M ratio, and SRT. The SBR process is unique as it is a batch flow process as opposed to continuous flow. Because of the batch nature of the process, it is typically conducted in multiple tanks. This approach allows one tank to be in fill mode accepting a new batch, while other tanks are in treatment mode. In smaller systems, it is possible to use one tank but a sufficiently sized equalization tank must be provided to store wastewater until the next batch can be fed to the process. The SBR process combines the dual function of an aeration tank and a secondary clarifier in a single tank.

The various treatment steps in an SBR typically proceed in four timed cycles called fill, react (or aerate), settle, and decant. It would be common for an SBR tank to undergo three or four complete cycles per day. Typically, at the start of the cycle, the tank is at its lowest liquid level, ready to accept a new batch. Wastewater is introduced during the fill phase until the high liquid level is achieved. The aeration system is turned on to introduce air to the process for BOD_5 oxidation. At the end of the aeration phase, the tank is allowed to settle before clarified liquid is drawn off as effluent during the decant cycle. Decanting is sometimes achieved using fixed decanters installed at a constant level in the tank; however, it is preferable to use a floating-style decanter that lowers automatically as the liquid level decreases. The floating-style decanter provides more process control flexibility. In some cases, plants may be equipped with a nonfloating decanter that is height adjustable. In a multitank SBR system, the phases of the cycle are offset between tanks to allow one tank to fill while another is in react mode.

As with virtually all aerated biological secondary treatment systems, there is always a potential for foaming. Foaming may not necessarily create a poor effluent quality but it will negatively impact the operation of the facility. At the very least, foaming is a nuisance that can consume significant operator time and detract from other more important process control activities that do impact the quality of the treated effluent. Considering the turbulence created by the diffused aeration system, the presence of some small amount of foam on the surface of the tank is expected. The degree of foaming can be significantly increased by the presence of filamentous bacteria. Typically, there are two types of filamentous bacteria called *Nocardia* or *Microthrix parvicella*. In extreme cases, foaming can overwhelm an aeration tank and expand over the walls of open-top tanks, creating a significant problem both for environmental concern and general plant housekeeping. Some treatment processes tend to be natural selectors that reduce the potential for growth of these microorganisms. Generally, the operator is in a reactionary mode with respect to foaming and must take quick action to reduce the impact of the event. Many plants are equipped with chemical dosing systems, such as chlorine as the dosing agent, to manage foaming events.

Status

Solids retention time is defined as the total sludge mass in the system divided by the rate at which solids are removed from the system, either via effluent TSS or WAS solids. The following equation 1 summarizes the calculation of SRT for a secondary treatment facility.

$$\text{SRT(d)} = \frac{\text{Total Sludge Mass (kg)}}{\text{Waste Sludge Mass (kg/d)} + \text{Effluent TSS Mass (kg/d)}} \qquad \text{EQ1}$$

In many cases, the mass of solids in the effluent is so small compared to the mass of solids in the WAS that it has little impact on the SRT calculation. To simplify the calculation, the effluent TSS mass can be assumed to be negligible at full secondary treatment plants achieving an effluent TSS concentration less than 25 mg/L. In these instances, a modified version of the standard SRT calculation shown below in Equation 2 can be used for process control purposes.

$$\text{SRT(d)} = \frac{\text{Total Sludge Mass (kg)}}{\text{Waste Sludge Mass (kg/d)}} \qquad \text{EQ2}$$

Rearranging Equation 2 to read as Equation 3 allows for calculation of the required WAS mass rate to achieve stable process control.

$$\text{Waste Sludge Mass (kg/d)} = \frac{\text{Total Sludge Mass (kg)}}{\text{SRT(d)}} \qquad \text{EQ3}$$

Controlling the secondary treatment system based on SRT as opposed to a target MLSS concentration may represent a philosophical shift in operational policy for many facilities. However, operations staff are encouraged to adopt this more comprehensive and scientific control strategy for increasing process stability. Using Equation 3 will enable operators to calculate the required WAS flow rate each day to maintain both a target MLSS concentration as well as a target SRT. The basic steps required to calculate the daily flow rate of WAS are as follows:

1. Quantify the mass of sludge in the aeration system based on the measured MLSS concentration. Quantify the mass of sludge in the secondary clarifier(s) based on the core taker suspended solids. Calculate the total

sludge mass by adding the mass of sludge in the aeration system and the mass of sludge in the secondary clarifier(s).

- ▲ Note that if a true TSS analysis cannot be performed at the plant, plant staff may use the Spin Test procedure, described in appendix A, to estimate TSS.

2. Select an operational target SRT, in days. The sludge mass control program, provided with this guide, helps to guide users through determination of SRT based on several basic input parameters.
3. Calculate the WAS mass rate required for stable process control using Equation 3.
4. Using the calculated WAS mass rate and the measured WAS MLSS concentration, calculate the WAS flow rate required.
5. These steps should be performed on a daily basis in order to achieve consistent and stable process control.

WAS is typically removed from the secondary clarifiers. The WAS operation and optimization is discussed further in the secondary clarifier section. From a stable process control perspective, it is important that the operator understands the impact of erratic wasting on the activated sludge process both within the aeration tank(s) and secondary clarifier(s).

The ultimate goal of the total sludge mass control concept is to ensure sufficient biomass inventory in the system to achieve the desired level of treatment. In the case of CAS, there needs to be sufficient biomass for BOD_5 removal. If nitrification is required, the quantity of biomass to achieve consistent nitrification will be higher. Estimation of the biomass required is commonly based on a food-to-microorganism (F:M) ratio for the facility. Essentially, this is the mass of BOD_5 (i.e., "food") being treated in the secondary treatment process divided by the total biomass. It is common to quantify the biomass using the mixed liquor volatile suspended solids concentration (MLVSS) multiplied by the total tank volume. It is common for a CAS process to operate at an F:M ratio in the range of 0.2–0.4 $kgBOD_5/kgMLVSS{\cdot}d$. The influent mass of BOD_5 is calculated by multiplying the concentration of BOD_5 by the daily flow rate. The target F:M ratio, combined with the influent mass of BOD_5, is used to solve for the mass of MLVSS required to achieve the target F:M ratio. At most facilities, the mass of MLVSS is converted to a mass of MLSS by factoring in the sludge volatility (ratio of MLVSS:MLSS, typically in the range of 70 to 80 percent). All of the parameters from these calculations should be tracked using trend charts to allow visualization of the data over time. The operators may conclude that, although they are operating consistently at the target F:M ratio and at the calculated total sludge mass, they are not achieving the desired level of treatment on a consistent basis. The trend charts will provide the data to support a decision to operate at a lower, more conservative F:M ratio. The F:M ratio can be lowered at a facility by increasing the biomass inventory (i.e., a higher MLSS concentration).

The plant OLR defines the mass of incoming BOD_5 per unit of aeration tank volume available. OLR is an additional monitoring parameter that is useful to track the secondary treatment system operation. Typically, systems are designed for a maximum OLR. This helps designers to quantify the volume of the aeration tank. Over time, it is common for the influent flow rate, BOD_5 concentration, or in some cases both parameters, to increase above the original design values. The result is that excess BOD_5 mass is being delivered to the aeration tank beyond what the plant was originally designed to handle. An overloaded plant such as this is likely to suffer in terms of treatment performance. There is little to nothing that the operator can do to directly impact the OLR in the aeration tank. The operator often has little control over the incoming flow and BOD_5 concentration, while the tank volumes are fixed. Tracking the OLR in a trend chart allows the operator to determine if excess loading may be the cause of performance issues at the facility. If the OLR is elevated, the operator can take some steps to respond to a high organic loading scenario in the aeration system as follows:

- As defined previously in this guide, the operator should monitor the BOD_5 removal performance in the primary clarifier. If the primary clarifier is Not Optimized, it could be the cause of the excess BOD_5 loading on the downstream aeration process.
- If the primary clarifier operation is optimized, the operator must adjust the operation of the aeration tank to cope with the higher OLR. Increasing the biomass inventory in the secondary treatment process may improve the performance of an overloaded system. Note that if the OLR is significantly higher than expected, the operating F:M ratio will be higher for a given biomass inventory. Increasing the inventory may reduce the operating F:M ratio to a more suitable target. Note: Practical limitations apply to how high the MLSS concentration can be increased before it negatively impacts the downstream secondary clarifier.
- It should be noted that more air will be required at higher operating OLR levels. If the aeration system is controlled via DO feedback, then this will be accounted for automatically as long as the aeration blowers have the required capacity.
- It may be possible to further enhance the removal of solids and BOD_5 in the primary clarifier through chemical addition. This is usually accomplished by the addition of a polymer or polymerized ballasted sand particles, which promotes better flocculation and therefore more efficient settling and removal in the primary clarifier. While this can work at some facilities to reduce the organic loading to the secondary treatment system,

the primary solids removal unit processes must be capable of handling the increased solids loading.

- In plants that are significantly overloaded, it may be possible that all of the optimization work described previously does not alleviate the problem. In these cases, the plant must either consider expansion (i.e., capital investment) or some means of reducing the influent loading at source. Reducing the loading at source can include requiring large commercial and industrial users to pretreat wastewater prior to discharging to the collection system.

In the case of a sequencing batch reactor (SBR) process, optimization of the system performance typically includes an analysis of the duration of the operating cycles. This can be quite difficult as many process variables need to be managed within a single tank. For example, under high flows and loadings, it may make sense to increase the size of the batch and extend the aeration phase to provide additional treatment capacity. This may work but any changes to the duration of one phase will impact the length of time available for the other phases. Extending the aeration phase may sacrifice the settle phase and result in unacceptably high suspended solids in the effluent. Any changes to either the fill or decant phase durations must consider the hydraulic capacity of the associated equipment. For example, if the decant cycle is shortened, do the decanters have the capability to remove the entire batch volume of clarified liquid in the shorter cycle? In general, it advisable to make small and incremental adjustments to the SBR cycles as it is very easy to inadvertently sacrifice one of the other phases.

The following questions are designed to help the operator better understand the optimization status of the CAS process. Operators are encouraged to develop and consider additional questions that may be specific to their utility:

- Have performance targets been developed for the secondary suspended growth treatment process? Recommended operating targets would include the following parameters:
  - ▲ Food-to-microorganism ratio (F:M).
  - ▲ Organic loading rate (OLR).
  - ▲ Target sludge mass inventory measured as total mass (lb or kg) of MLSS in both the aeration tank and the secondary clarifier.
  - ▲ Target residual dissolved oxygen (DO) concentration in the aeration tank.
- Does the plant's monitoring and testing program allow for sufficient data collection to thoroughly monitor actual plant performance versus the defined performance targets? Recommended testing and data collection for the aeration tank would include the following parameters. (Note that

additional related parameters, which may be applicable, are included in the secondary clarification and nitrification sections of this guide.)

  - Aeration tank influent BOD_5 concentration. (In many plants, this would be equal to the primary clarifier effluent BOD_5 concentration and there would be no need to test again if this has already been captured.)
  - Aeration tank MLSS and MLVSS concentrations.
  - Aeration tank DO concentration. (At many plants, these data are captured and logged automatically using online probes and SCADA software.)

- Are the appropriate trend charts being developed and populated to capture all of the raw data and calculated performance parameters to properly monitor the aeration tank?
- In the case of an SBR system, the following questions apply:
  - Have the appropriate cycle and phase durations been identified and optimized for actual plant operation?
  - If the plant has a nonfixed style decanter (i.e., floating or adjustable), has the operation been optimized?
  - Has the batch volume per cycle been optimized to suit the actual operation and number of cycles per tank?
- Is the plant equipped with an appropriate chemical dosing system or other means to combat foaming or filamentous events?
- Has a regular maintenance and inspection program been developed for the plant that includes daily inspection of the aeration tank and associated equipment?
- Have SOPs been developed for the aeration tank, including an emergency response plan?
  - The emergency response plan should also include specific procedures to allow quick response to a foaming event.

Process Energy Considerations

At many treatment plants, the secondary treatment process is where the largest energy consumption takes place. The aeration system typically is the single largest energy consuming unit process by a wide margin. Dissolving oxygen into the aeration tank is an absolute requirement for the biological process; however, it is a comparatively inefficient process. Depending on the site conditions and other operating parameters, the oxygen transfer efficiency of a diffused aeration system is often less than 35 percent. Regardless of the efficiency of the equipment used, the physical conditions and limitations (i.e., temperature, altitude, etc.) make it theoretically impossible to achieve higher levels of oxygen transfer efficiency for a specific site. The energy efficiency considerations for the aeration process, therefore, focus heavily on the control of the aeration system based on variable oxygen demand within the system.

For an activated sludge process, a theoretical mass of oxygen is required to complete the BOD removal. A theoretical mass of oxygen is required to achieve full nitrification as well. As the influent loading of either BOD or TKN changes, the corresponding demand for oxygen will also change. In addition, as site conditions change (most notably, temperature), different volumes of air input are required to deliver the same mass of oxygen to the process. While these processes are complex from a chemistry perspective, they are relatively easy to monitor. Many plants are equipped with DO sensors that monitor the oxygen concentration in the mixed liquor in the aeration zone. Often, the aeration blowers have variable speed capability and DO monitoring can be used to automatically adjust the blower speed to meet the process oxygen demand. When flow rates and organic loadings are lower, this results in the blowers running at reduced speed, and results in significant energy savings.

Often, aeration tanks operate at a target DO concentration of 2 mg/L. By maintaining a residual of 2 mg/L, the aeration system is adding 2 mg/L more oxygen than what is actually required to achieve biological treatment. This approach is often applied as a precaution to ensure enough oxygen has been added to achieve treatment, with an additional buffer. Maintaining any residual DO in the aeration tank represents excess dissolved oxygen that is not required for treatment. Opportunity exists to find efficiencies and energy savings. For example, it may be possible for operating staff to reduce the target residual DO. In some case studies, plants are operating at target DO concentrations of 0.5 mg/L or 1.0 mg/L. This may not seem like a major change; however, depending on the size of the facility, any reduction in air flow rate and blower speed is a savings in energy use. The magnitude of the savings will be highly dependent on the unique conditions at each facility.

At facilities that nitrify, it may be possible to use additional process monitoring to add a higher level of process control for the aeration system. The reactions for the nitrification process tend to proceed more slowly than the BOD oxidation reactions. Therefore, if the system is removing all of the ammonia through nitrification, it can be reasonably assured that the BOD_5 has also been completely removed. If the plant happens to be equipped with online ammonia probes in the aeration tank(s), it may be possible to use them as an additional layer of instrumentation control. If the ammonia is being completely treated, sufficient oxygen has been added to treat both BOD_5 and ammonia. Some plants may use the online ammonia reading as a means to control the aeration blowers. This type of control would tend to minimize the DO residual in the aeration tank while still ensuring that sufficient oxygen has been added to achieve full nitrification. This approach is not currently a widespread control strategy in the industry, largely due to the relatively high cost of online ammonia instrumentation. However, for larger facilities that may already have this instrumentation in place, there may be opportunities to use the approach to achieve a very high level of aeration system control.

The main goal of the treatment facility is to ensure the proper treatment of wastewater to the required effluent quality standards for protection of public health and the environment. The demands for air in the secondary treatment process must be met at all times to ensure the treatment process is operating properly and is maintaining a viable activated sludge biomass. A minimum amount oxygen must be delivered and there are practical limitations as to how efficiently this oxygen can be delivered using the diffused aeration system. The delivery of this oxygen at most facilities represents a large energy consumption compared with other unit processes. Any savings in the aeration system can represent significant energy and operating cost savings.

To assess the optimization status of energy efficiency with respect to the CAS process, plant staff are encouraged to consider the following questions:

- Is the current aeration diffuser system providing acceptable oxygen transfer efficiency?
- Does the plant use dissolved oxygen monitoring for blower speed control?
- Is the current aeration blower technology providing the best efficiency?

In summary, utility staff are encouraged to consider the following modifications and/or upgrades that could result in potentially significant energy savings. A cost analysis is recommended to compare any capital investment required for the upgrades versus the life-cycle cost savings due to reduced energy consumption. Typically, a present value analysis would be performed, including calculation of the payback period to determine the long-term impact of the upgrade.

- Investigate the existing aeration tank diffuser system. If the diffuser system is aging and/or if it is based on coarse bubble aeration, replacement with a new fine bubble diffuser system may offer better oxygen transfer efficiency. Fine bubbles are more efficient for oxygen transfer due to the larger surface area to volume ratio of the bubble. Any improvement in oxygen transfer efficiency means less air flow is required to deliver the same mass of oxygen to the process.
- Dissolved oxygen instrumentation coupled with speed control for the aeration blowers. Investigate the existing aeration blowers to determine if the electric motors are compatible with variable frequency drive (VFDs) operation. If online dissolved oxygen instrumentation is not available at the plant, it may be worthwhile to add this instrumentation both for better process control and energy savings. The general concept is to control

the speed of the aeration blowers to match the oxygen demand of the system. Further efficiencies can be found by optimizing the target residual DO concentration (consider using 1.0 mg/L or 0.5 mg/L). The potential energy cost savings that may be derived through the use of automated DO control can far exceed the initial costs of instrumentation and controls.

- If online ammonia instrumentation is already in place at the facility to monitor the nitrification process, it may be possible to use this instrumentation to add a further level of control to the aeration blowers. Due to the relatively high cost of this instrumentation, it is often not worthwhile to add this to the plant for the sole purpose of saving energy, particularly if the facility already uses DO instrumentation for blower speed control.
- Plant staff should investigate the existing aeration blowers to determine the technology and age of the equipment. Advancements in blower technology can result in significant improvements in the energy efficiency of the technology. Often, energy savings can be realized even through replacement of an existing unit with a more modern version of the same technology.

Action

If areas of the CAS process operation are considered Partially Optimized or Not Optimized, the following steps are recommended for implementing an action plan.

- Treatment performance objectives should be developed and clearly defined for the aeration tank operation. If objectives have not already been developed, operators are encouraged to consult existing design documentation for original design operation targets. If no design information is available, typical operating guidelines for various activated sludge processes are published in numerous texts. For a CAS process, typical performance objectives would be as follows:
  - F:M Ratio: 0.2–0.4 kg BOD_5/kg MLVSS·d
  - OLR: 0.3–0.7 kg BOD_5/m3·d
  - SRT: 5–10 days for BOD_5 removal only (i.e., no nitrification)
  - Dissolved oxygen: 2.0 mg/L (consider targeting 0.5 or 1.0 if optimizing for energy)
- Develop and implement a regular sampling, testing, and monitoring program for the aeration tank. It is recommended that the program be implemented on a daily basis for most facilities. Minimum data collection should include the following parameters:
  - Aeration tank influent BOD_5 concentration.
  - Aeration tank MLSS and MLVSS concentrations.
  - Aeration tank DO concentration.

- Implement the use of trend charts for tracking all data collected for the facility, including the aeration tank. Trend charts are a beneficial tool that staff should update and review on a daily basis for process monitoring. Trend charts for performance targets should display the target to provide the operators with a clear visual check of how actual performance compares with the defined objectives.
- In the case of an SBR, steps may be taken to adjust the cycles per day and the duration of the individual phases within each tank to optimize the various processes. Care should be taken when making any adjustments to either the fill or decant cycles to ensure the plant equipment (e.g., influent pumps and decanters) has the required hydraulic capacity. Settling tests will also be vital if any adjustments are made to shorten the settling phase duration. If the plant is experiencing poor effluent solids concentrations, it may be necessary to increase the settling time, which impacts all of the other phases. In general, it is recommended to make slow and small adjustments to the SBR cycle and phase times so as not to cause a process upset.
- A regular maintenance and visual inspection program should be developed for the entire facility, which includes the aeration tank. The operator should make sure to visually inspect the aeration system on a daily basis for the following potential issues:
  - Foam: Note that some plants may have some manageable foam or scum layer on the surface of the tank. It is more important to take note of any sudden increases in the foam layer that may indicate the presence of filamentous bacteria. It can be helpful to microscopically observe the organisms present in the aeration basin to further discern the nature of the biomass present.
  - Even distribution of aeration bubbles across the surface of the tank. Uneven distribution may indicate a blockage in an aeration diffuser module, a crack in a main aeration header, or a potential issue with one of the aeration blowers.
  - In the case of an SBR with movable decanters (either adjustable or floating style), regular inspection should ensure that the decanter mechanism is not being hindered in any way and is free to move as needed. In cold weather climates, the accumulation of ice is common and can prevent proper movement.
- SOPs for the aeration tank should be developed, including an emergency response plan and maintenance procedures. For the aeration tank, it is specifically recommended that the emergency response plan include procedures for dealing with a significant foaming event. The operator along with plant management are encouraged to take an active

role in developing SOPs. The operator should regularly review the procedures and the emergency response plan to be better prepared in the event of an unexpected issue with the system. The operations team is encouraged to continually update and improve SOPs and response plans, and to provide training to all pertinent staff. Well-documented SOPs will provide better continuity in the event of staff changes within the facility.

Attached Growth

Understanding

An alternative to the suspended growth biological treatment process is the development of a biofilm layer that grows on specially designed media. Attached growth processes can be subdivided into two main categories: submerged and nonsubmerged. Submerged attached growth processes use a combination of media-based biofilm growth submerged in a suspended growth aeration tank. These types of systems include a wide range of media packing materials and designs. The media can be attached or floating. In attached systems, the packing material is permanently attached to a support structure submerged in the activated sludge. In floating systems, smaller packing is allowed to float freely in the bioreactor, and it uses an effluent screen to retain the media in the aeration tank. These types of systems typically rely on diffused aeration similar to the CAS process for treatment. The approach to operating submerged attached growth processes can be similar to suspended growth treatment processes, including many of the process control techniques.

Rotating biological contactors (RBC) are another variation on the submerged attached process where large media disks are partially submerged. RBC processes rely on the rotation of the disks to provide aeration by natural diffusion as the biomass is exposed near the top of disks that extend above the liquid level.

Nonsubmerged attached growth processes use a biofilm layer grown on specialized media that is not inserted into a suspended growth aeration tank. An example of this type of process is a trickling filter. Typically, a trickling filter is packed with media (commonly rocks) that is specially sized to allow for a high packing density while still having void space throughout the bed. It is common for trickling filter media to be contained within a circular concrete tank. Raw wastewater is evenly distributed using a rotating spray bar over the top of the media and allowed to percolate through the media bed. Conditions within the trickling filter favor the growth of a biofilm on the surface of the media that treats the wastewater as it passes through the media bed. Aeration is accomplished by natural diffusion of air as the wastewater passes through the void spaces in the media bed.

Status

Submerged attached growth treatment processes operate similar to suspended growth processes in that aerobic bacteria in the aerated media tank break down soluble BOD_5 and ammonia in the wastewater. The mixed liquor then passes through to a solids separation process such as a secondary clarifier. In some cases, submerged media reactors can even be followed by MBRs for a higher level of solids removal. The system functions similarly to suspended growth processes, including many of the process control techniques. For example, a sufficient mass of activated sludge must be maintained in the system to achieve the desired level of treatment. Once the target mass of activated sludge is calculated, it can be monitored and tracked using the total sludge mass control program to ensure that the process control is stable. The submerged media's design provides additional surface area for the biomass to become attached. The combination of biomass in suspension and attached to the media allows for a higher biomass to be maintained in the same reactor volume.

Measuring the MLSS concentration can help manage the sludge mass in submerged attached growth systems. Measuring the concentration of MLSS in the aeration tank, however, will underestimate the total mass in the system. This is because biomass attached to the media will not be included in the MLSS sample. In a well-operated system, some natural sloughing of biomass from the media will occur, as well as regrowth of new biomass on the media, that will establish an equilibrium. Therefore, it is practical to use the activated sludge mass in suspension and compare it with the target value to calculate how much sludge mass to waste per day.

A starting point for applying the total sludge mass control concept to submerged attached growth processes is to select a target SRT using typical guidelines to calculate the target sludge mass in suspension. With practice, the operator will be able to zero in on the optimal target sludge mass once an equilibrium between the biomass in suspension and attached to the media is established and stabilized.

The mass control program can be adapted for application to RBC processes but it can be a challenge to ensure aeration of the entire suspended growth biomass, particularly at higher OLRs. This can result in anaerobic zones within the process that can lead to septicity and odor issues. This makes it even more challenging for the operator in the event that additional biomass is required in accordance with the mass control program. It may not be possible to sustain the higher biomass if limitations in the process design do not allow for sufficient aeration.

Trickling filters are limited in terms of optimization as it is not possible to quantify the biomass in the system using normal sampling techniques. Therefore, this type of process cannot be controlled using a target SRT or the total sludge mass control program. Alternative techniques for operation and control must be applied. The operator is limited to operating the trickling filter in accordance with typical organic and volumetric loading guidelines. The operator is also limited on process control modifications that can be made to adjust to higher-than-normal loadings that exceed the design limitations of the system. Similar to the RBC process, high organic loading can lead to septic conditions and odors, and the operator has few options at their disposal to respond to these issues.

For both the RBC process and the trickling filter, assessing process capacity may be more practical than optimization. If the operator is struggling with the process, the best approach is to maintain diligent sampling and testing records as evidence of due diligence as well as collect adequate data to support upgrading the system.

The following questions are designed to help the operator better understand the optimization status of attached growth processes. Operators are encouraged to develop and consider additional questions that may be specific to their utility:

- Have performance targets been defined for the process? Performance targets are recommended to include the following parameters:
  - The attached growth process is capable of removing soluble BOD_5 and TSS similar to the suspended growth CAS process. Depending on the design of the system, the attached growth process may also be capable of nutrient removal. Performance targets should include all effluent discharge limits for the system.
  - Target SRT and total sludge mass should be defined in accordance with the total sludge mass control program introduced in the suspended growth section of this chapter and further discussed in the secondary clarification section. Selecting a target SRT is the first step and should include all treatment process objectives, including nutrient removal.
- Has a sampling and testing program been developed for the attached growth process that allows actual performance to be monitored compared to the defined targets?
  - Operators are encouraged to explore the suspended growth and nutrient removal sections of this chapter for recommendations to develop a sampling and testing program.
  - The sampling and testing program should also include all parameters required to apply the sludge mass control program (with the exception of trickling filter processes).
  - This would include testing of the media tank and any downstream secondary clarifiers.
- Has the collected data been plotted on trend charts along with any applicable performance targets?
  - Trend charts should be updated and reviewed daily for signs of reduced performance.

  - Daily review of trend charts, particularly the sludge mass control program, is required to minimize deviations from the selected performance targets.
  - The data can also be used to help operating staff make the correct process control adjustments.
- Does the plant have the capability to provide sufficient aeration to meet the treatment requirements? This is of particular concern for processes such as RBCs or trickling filters where aeration is accomplished by natural diffusion.
- Does the attached growth process require regular replacement of the media and does the plant have sufficient spare media on-site?
- Has a regular inspection and maintenance program been developed for the attached growth process?
  - Recommendations for inspection would include items defined in the suspended growth and nutrient removal sections of this guide.
  - Additional items would include visual inspection of the media for biofilm development and any signs of damage that may require replacement.
  - Effluent screens should be regularly inspected to ensure they are not clogged with debris and are retaining the floating media in the bioreactor.
- Have SOPs been developed for the attached growth process, including an emergency response plan?

Process Energy Considerations

With the exception of the media, attached growth treatment processes are very similar to suspended growth systems. This is especially true from a process energy perspective where aeration is still the largest consumer of energy. Operators are encouraged to review the applicable suspended growth and nutrient removal process energy sections of this chapter for additional information.

Action

If areas of the attached growth process operation are considered Partially Optimized or Not Optimized, the following steps are recommended for implementing an action plan.

- Performance targets should be clearly defined for the process to meet the requirements of the effluent discharge permit and the total sludge mass control program. The operators should consult the operating permit as well as other applicable sections of this guide to help develop the required performance targets.

- A sampling and testing program should be developed that allows for tracking of actual performance relative to the defined performance targets, including the requirements of the total sludge mass control program if applicable (i.e., does not apply to the trickling filter process). Sampling and testing should be completed on a daily basis, particularly for application of the mass control program.
- The plant should complete a thorough review of aeration capabilities, especially for trickling filters and RBC processes, in accordance with the process capacity assessment guidance included in chapter 3. These processes do not rely on mechanical aeration and practical limitations apply to the amount of air that can be provided. This will impact the maximum OLR and biomass that can be maintained within the system. Implementation of a good sampling, testing, and data monitoring program will assist with the evaluation. If the system is incapable of providing sufficient air for the RBC or trickling filter system, it is likely the only recourse will be to upgrade the process.
- A regular inspection and maintenance program should be implemented for the plant. Daily inspection for aeration, mixing, and foaming should be included. Regular inspection of media for biofilm development and any effluent screens should be included.
- SOPs should be developed for the system, including an environmental management plan. The original design documentation for the attached growth process should provide information to assist with developing appropriate procedures for media handling and maintenance.

Secondary Clarification and Total Mass Control Concept Understanding

Secondary clarifiers are typically constructed in either a circular or rectangular tank configuration. Similar to primary clarifiers, the separation of solids is accomplished by gravity. Proper mechanical operating conditions need to be maintained in order to ensure good settling. Secondary clarification and sludge settling are highly dependent on the surface area, depth, and specific operating conditions of the clarifier basin. The operator has no control over the dimensions of the clarifier but can control the operating conditions.

Two key operational parameters for the operation of secondary clarifiers are the surface overflow rate and the solids loading rate. Surface overflow rate is calculated as the flow rate through the clarifier divided by the total surface area. Solids loading rate is defined as the mass of solids fed to the clarifier divided by the total surface area. The secondary clarifier is usually designed to operate within typical ranges for both of these parameters. Daily calculation and monitoring of these operating parameters may assist staff with optimizing clarifier operation, especially in systems with an opportunity to bring standby clarifiers into operation. Excessive solids loading to the clarifier may also be an indicator that too much biomass inventory is being carried in the aeration tank, or possibly the primary clarifier is not removing enough solids prior to the secondary treatment process.

Secondary clarifiers are usually equipped with inlet baffle structures to help regulate inlet velocities and prevent short circuiting, surface skimmers to remove any scum buildup, and an effluent weir structure to allow overflow of clarified effluent. Similar to primary clarifiers, weirs are typically linear for rectangular clarifiers and circumferential for round clarifiers.

The primary operating philosophy of the secondary treatment process is to maintain an active biomass in the aeration tanks that is sufficient to biodegrade the incoming BOD_5 and, in some facilities, to support nitrification for ammonia conversion. In many cases, the MLSS concentration is in excess of 3,000 mg/L to provide the biomass needed for these processes. The operators are encouraged to view the biomass as a valuable commodity that needs to be scientifically managed. The purpose of the secondary clarifier is to allow the MLSS to be separated by gravity settling. This separation is designed to reduce the TSS in the effluent to low levels to meet the effluent discharge requirements. Often, this is the only function of the secondary clarifier that operators give significant thought to regarding process control and optimization.

While it is important to ensure the effluent TSS criteria are met, it is equally important to consider the impact of the clarifier operation on the upstream aeration tank operation. The secondary clarifier settles suspended solids. It is important to view these solids as valuable organisms that make up the total system biomass. The material collected in the clarifier is returned to the aeration tank(s) via RAS pumping systems to sustain the treatment process. Managing the RAS system is a complex balance between maintaining an adequate quantity of biomass in the aeration basin, while at the same time minimizing the quantity of biomass in the secondary clarifier. BOD_5 oxidation and nitrification occur in the aeration basin only, and control of the RAS flow rate needs to be carefully monitored to avoid excessive buildup of sludge in the secondary clarifier. Conversely, the RAS flow rate needs to allow for some accumulation of sludge in the clarifier to ensure that properly thickened sludge is being returned instead of just water. These two objectives for control of the RAS system may seem counterproductive. That is why the operation of the system must achieve a delicate balance to optimize both functions simultaneously through careful monitoring and observation. Controlling the RAS and WAS flows to optimize the biomass distribution and concentration between both the aeration tank(s) and the secondary clarifier is the basis of the total sludge mass control concept. A diagram of RAS and WAS flow is displayed in Figure 4-7.

Status

The basic operation of the secondary clarifier is similar in concept to primary clarification. The objective is to remove suspended solids material from the wastewater flow before it is discharged over the outlet weir. The secondary clarifier has a unique set of process goals. From a treated effluent quality objective, the clarifier needs to ensure that enough suspended solids are settled out to produce an effluent quality that meets the effluent discharge requirement. In plants with no tertiary treatment, the secondary clarifier is often the final barrier to excess solids being discharged to the receiving environment. A number of operational parameters will help monitor the performance of the clarifiers, including influent solids loading rate and the surface overflow rate. The design documentation for the facility should define the operational ranges that the clarifiers were designed for. If no design documentation is available, many resources provide published guidelines for typical operating ranges. In keeping with the theme of trend charts, the operator is encouraged to monitor the suspended solids entering the clarifier along with the flow rate to be able to calculate both the solids loading rate and the surface overflow rate. These calculated values should be plotted on trend charts, along with the high and low operating range limits, to provide a quick visual confirmation of the clarifier operation.

One of the most critical parameters impacting the performance of the secondary clarifier is the characteristics of the sludge itself. The operator should be particularly aware of creating sludge settling characteristics that contribute to poor settling sludge. One of the simplest ways to monitor the sludge properties is to complete a sludge-settling test on a regular basis. A sample of sludge is

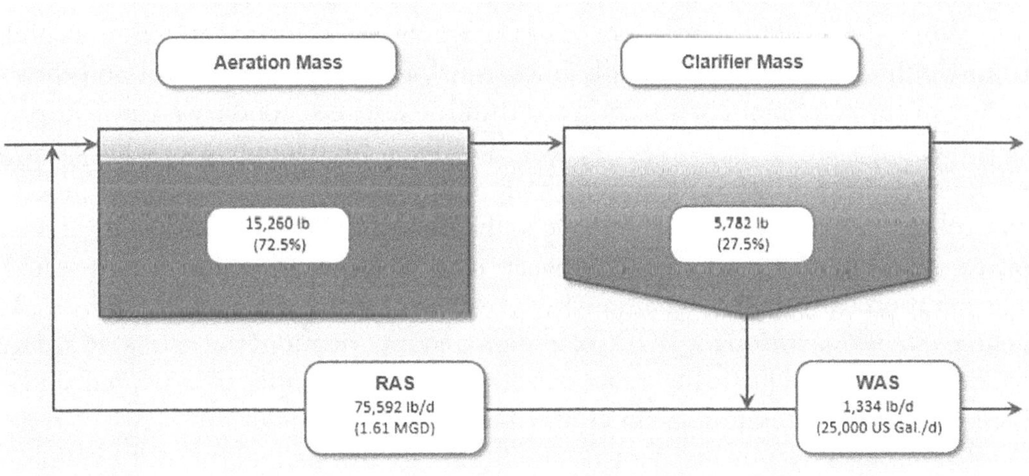

(Source: Daryl Burke, EXP Services, Inc.)

Figure 4-7. Illustration of the location of RAS and WAS flows

collected from the aeration tank and allowed to settle by gravity in a beaker or settleometer. To quantify the capability of the sludge to settle, the volume of settled sludge is recorded after a set time period for each test. Poor settling sludge will result in a higher volume or possibly a scenario in which the sludge does not settle at all in the bench test. The operator can record the settled sludge volume on a trend chart to observe changes over time. It is also helpful to plot effluent TSS on this chart so that effluent TSS concentrations can be correlated with the settling test results.

An additional concern for secondary clarifiers is the presence of bulking or filamentous sludge that leads to poor settling. Filamentous bacteria interfere with the formation of a dense floc in the clarifier that contributes to reduced settling performance. The operator should have the capability to observe and identify filamentous bacterial cells with the use of a microscope. The operator may need to make process control adjustments or begin adding chlorine to decrease the filamentous bacterial population. Other process control conditions that can promote filamentous growth include: low dissolved oxygen in the aeration tank, high RAS recycle rates, and septic conditions in the raw wastewater/headworks. Plant staff can make incremental adjustments to improve these conditions, which may reduce the impact of filamentous growth. In emergency situations or where the above process control parameters have already been optimized as much as possible, the operator should have the capability to dose chlorine to the system to combat filamentous growth. Chlorine is typically dosed in the RAS line. The dosage must be scientifically applied to target the filamentous bacteria and minimize widespread impact on the viability of the biomass.

In addition to measured parameters, the secondary clarifier can be prone to operating conditions that are difficult to quantify and require regular observation of the process. The operator should check the clarifiers on a daily basis for excess scum buildup, check for excess debris or biofilm on the outlet weirs, and ensure that skimmers and other mechanical equipment are operating properly.

While the most obvious purpose of the secondary clarifier is to remove solids from the final effluent, an equally important purpose is the control and distribution of the total biomass inventory within the entire secondary treatment process (i.e., aeration tanks plus secondary clarifiers). Managing suspended growth systems requires utility staff to implement and maintain a stable level of process control. Stable process control is especially important where utility staff choose to, or are required to, achieve a higher level of treatment (such as nitrification). The aeration basin is the dynamo where treatment is achieved and sustained. To achieve stable operation for BOD_5 oxidation and nitrification, the activated sludge barrier must be scientifically managed. An adequate quantity of activated sludge is required in the aeration basin at all times to achieve the desired level of treatment. Establishing process control conditions in the secondary unit processes that result in stable operation is a complex balance between carrying an adequate quantity of activated sludge, while at the same time minimizing the mass of activated sludge in the secondary clarifier. Traditional process control programs tend to focus on the concentration of mixed liquor suspended solids (MLSS) as the primary process control parameter. The amount of activated sludge to waste is estimated, and the general approach is to maintain a target MLSS concentration in the aeration tank. It is rare for facilities to measure the quantity of activated sludge in the secondary clarifier. Significant amounts of activated sludge can be present in the secondary clarifier, and they will continue to accumulate if the RAS flow rate is not scientifically managed by the operating staff to ensure the quantity of activated sludge is minimized. Operators are encouraged to measure the height of the sludge layer in the secondary clarifier as well as the MLSS concentration of the settled sludge to help determine the mass of sludge in the clarifier. At many facilities, it may be much easier to sample the RAS flow for solids concentration, which gives a usable approximation of the sludge concentration that will be wasted from the bottom of the secondary clarifier.

Figure 4-8 illustrates the total sludge mass control concept. Implementing the total sludge mass control concept is a more comprehensive and scientific approach to managing the activated sludge barrier. The approach involves measuring the mass of activated sludge both in the aeration tank and the secondary clarifier. The sum of the aeration tank mass and the secondary clarifier mass equals the total mass of activated sludge in the secondary unit processes. Using the solids retention time (SRT) formula shown in Figure 4-8, operating staff calculate the mass of sludge to waste. The mass of activated sludge to waste is calculated by dividing the total sludge mass by the target SRT. A spreadsheet used to measure and track the total sludge mass is provided to Partnership for Clean Water program subscribers.

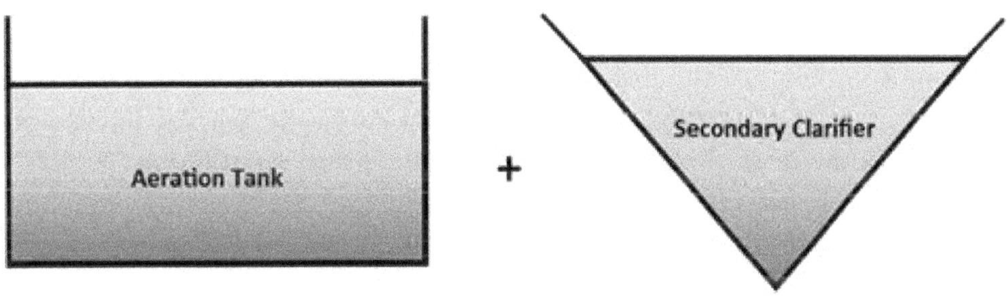

Total Sludge Mass (kg) = Aeration Sludge Mass (kg) + Clarifier Sludge Mass (kg)

$$\text{SRT (days)} = \frac{\text{Total Sludge Mass (kg) (Aeration + Clarifier)}}{\text{Sludge Mass Wasted (kg/day)}}$$

(Source: Gerard Wheeler, Global Facilitation, Inc.)

Figure 4-8. Total sludge mass control concept

Case Study: Total Sludge Mass Control Concept Application

A trend chart that illustrates the application of the total sludge mass control concept is provided in Figure 4-9. The WWTP shown is an approximately 100 MGD CAS facility designed for both BOD_5 oxidation and nitrification (note that nitrification is described in more detail in the nutrient removal section of this chapter). The top-most horizontal trace represents the target sludge mass that was required to support consistent nitrification. The bottom line on the chart represents the seven-day moving average of the mass of activated sludge in the secondary clarifiers. The middle line represents the seven-day moving average of the mass of activated sludge in the aeration basins. The top-most line represents the seven-day moving average of the total sludge mass (aeration plus secondary clarifiers). The task for the operators responsible for process control is to maintain the actual total sludge mass as close as possible to the target. Keeping the actual sludge mass close to the target ensures an adequate quantity of nitrifying bacteria to achieve consistent treatment performance (nitrification in this case). The trend chart confirms the operators did maintain the total sludge mass close to the target.

The improvement in nitrification achieved by maintaining the actual sludge mass close to the target is illustrated in Figure 4-10. The period evaluated in Figure 4-10 is September 2009 to May 2012, and it reflects the two liquid trains employed at the plant, Plant 1 and Plant 2. The effluent ammonia concentrations for Plant 1 and Plant 2 are represented by the labeled lines on the trend chart. The final effluent seasonal target for ammonia is depicted by the line on the graph labeled Ammonia Target. Between January and March 2010, the final effluent ammonia concentration from both liquid trains ranged from 8–10 mg/L, higher than the final effluent ammonia target of 5 mg/L. The same pattern was repeated between January and March 2011. Prior to July 2011, the approach to process control was to maintain only the concentration of mixed liquor suspended solids (MLSS) in the aeration tank. There was no measurement or knowledge of the mass of activated sludge in the secondary clarifiers.

In response, city staff elected to implement a Composite Correction Program, which incorporated the total sludge mass control concept, to identify and resolve performance limiting factors in July 2011 (see Figure 4-10). This effort also established stable process control in advance of the approaching cold winter season. Between August and December 2011, multiple administration and operations limiting factors were resolved such that the activated sludge barrier was fully optimized as the winter season approached. Following the implementation of the total sludge mass control concept, operating staff achieved a level of nitrification with the existing unit processes that was unprecedented. For the first time, stable process control was maintained by the operating staff, between January and March 2012. As a result, the level of nitrification achieved was significantly higher compared with the two previous winter periods. Between January and March 2012, the final effluent ammonia concentration from both liquid trains was below 2 mg/L and below the seasonal final effluent ammonia target of 5 mg/L.

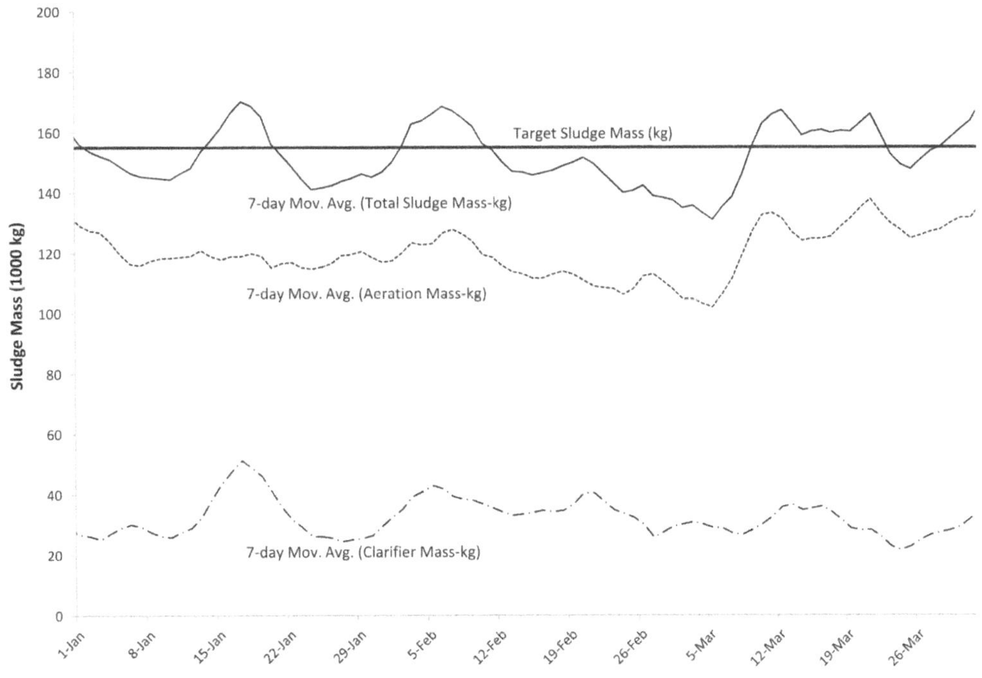

(Source: Gerard Wheeler, Global Facilitation, Inc.)

Figure 4-9. Total sludge mass control trend chart

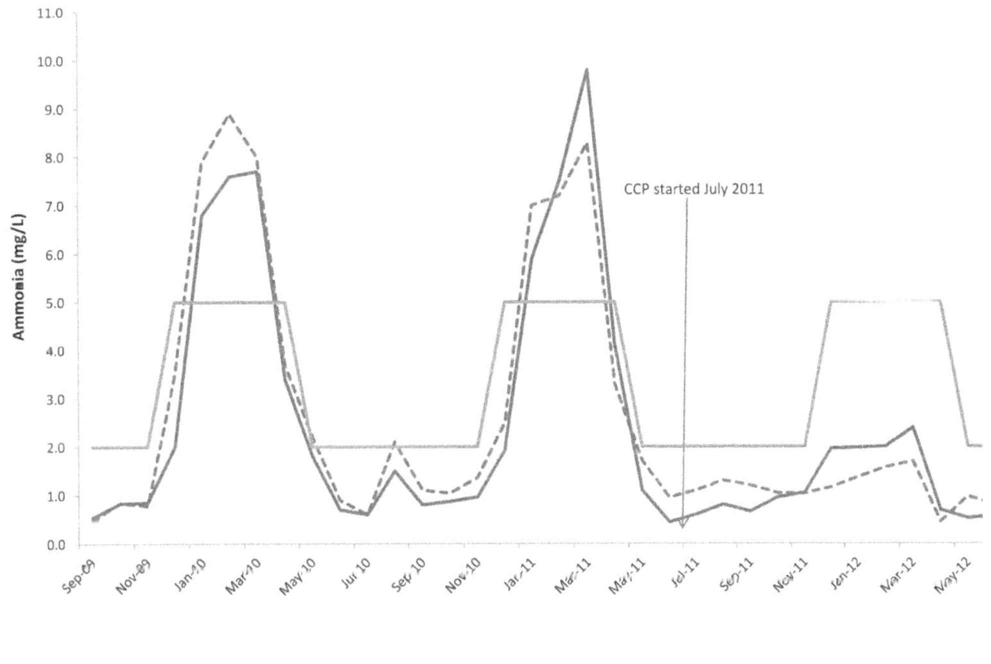

(Source: Gerard Wheeler, Global Facilitation, Inc.)

Figure 4-10. Final effluent ammonia concentration

The following questions are designed to help the operator better understand the optimization status of the secondary clarification process. Operators are encouraged to develop and consider additional questions that may be specific to their utility:

- Have performance targets been developed for the secondary clarifier? Key performance targets for the secondary clarifier should include the following:
  - Clarifier effluent TSS concentration. At facilities without tertiary treatment, the clarifier effluent TSS target would be the same as the plant effluent TSS limit in accordance with the operating permit. Developing targets with a margin of safety with respect to the plant permit limits is recommended.
  - Operating ranges for solids loading rate and surface overflow rate.
  - Target biomass inventory (aeration plus secondary clarifier) and sludge blanket level in the secondary clarifier.
  - Target TSS concentration of the settled solids/RAS.
- Has a sampling, testing, and monitoring program been put into place for the secondary clarifier? The monitoring program should include monitoring of parameters required for the total sludge mass control concept introduced in this guide. A summary of the recommended parameters is as follows:
  - Clarifier influent flow rate.
  - Clarifier influent TSS concentration (may be the same as the aeration tank MLSS concentration).
  - Clarifier effluent TSS concentration.
  - Settled sludge level. This parameter is used to estimate the volume of settled sludge in the clarifier.
  - Settled sludge TSS concentration. In most cases, this value will be more easily obtained by sampling from the RAS piping.
  - Observation of mixed liquor sludge samples under a microscope for the presence of filamentous bacteria.
- Has a sampling and monitoring program been developed to monitor the settling characteristics of the sludge? This should include, as a minimum, regular sludge-settling jar tests and recording of the settled sludge volume. Additional testing can be performed as needed by external laboratories to help identify the presence of certain bacterial strains or other properties of the sludge that may be leading to poor settling.

- Have the appropriate trend charts been developed to properly track the collected data and monitor the clarifier performance relative to the defined operating targets?
- Has the operation of the RAS system been developed to optimize the distribution of biomass between the aeration tank(s) and the secondary clarifier and to satisfy the goals of the total sludge mass control concept? In general, this is a balancing act that requires careful monitoring and control to ensure sludge thickens sufficiently in the clarifier, and is transported back to the aeration tank(s) to maintain the maximum possible quantity of activated sludge under aeration.
- Has the sludge wasting program been developed on the basis of the total sludge mass control concept that includes operating at a target SRT?
- Have steps been taken to minimize filamentous bacterial growth that leads to poor settling? Steps should be taken to optimize the following conditions in order to minimize filamentous bacterial growth:
  - ▲ Minimize retention times in raw wastewater holding tanks, pump stations, and headworks equipment to minimize potential for septic conditions.
  - ▲ Ensure sufficient aeration is being provided to the biomass in the aeration tank at all times. In the absence of DO feedback for blower control, adding more air to the process may reduce filamentous growth.
  - ▲ Reduce RAS or other internal recycle rates as long as it does not otherwise compromise the treatment performance.
- Does the operator have the ability to dose chlorine for control of filamentous bacterial growth? If process control techniques do not sufficiently control filamentous growth, chemical addition may be required.
- Has a maintenance and inspection program been developed for the secondary clarifier? Inspection items should include monitoring of the scum buildup, checking the effluent weirs for buildup or biofilm, and observing the operation of mechanical systems such as scum skimmers.
- Have SOPs been developed and documented for the secondary clarifiers, including an emergency response plan?

Process Energy Considerations

The secondary clarification process is largely a passive process that relies on gravity to settle the activated sludge in secondary clarifiers. There are some related processes that do consume energy and may represent an opportunity for improved efficiency. The main related processes are the return activated sludge (RAS) and waste activated sludge (WAS) systems. These two systems generally rely on the use of mechanical pumping and, in some cases, depending on the plant hydraulics, the RAS pumps can be used to generate the WAS flow. In this case, the WAS is typically removed through the use of a valve-controlled pipe attached to the RAS piping. The following opportunities for improved efficiency should be considered.

- Any reduction in the speed of the RAS pumps will represent an energy savings to the plant, provided it does not impair the treatment process performance.
  - ▲ The primary objective of the RAS flow is to recycle settled biosolids from the secondary clarifier back to the aeration tank. The flow rate needs to be adequate to maintain the majority of the biomass in the aeration tank in accordance with the total sludge mass control concept. The specific rate of return solids required is highly dependent upon the mass rate of biosolids entering the clarifier. In many plant designs, it is common to include flow paced speed control on the RAS pumps based on the raw wastewater influent flow rate. This allows for the RAS pump speed to be proportionately reduced during periods of low flow to the plant, and similarly increased during periods of high flow to the plant. In a typical activated sludge process, the RAS flow required is in the range of 0.5 to five times the influent flow rate, depending on the specific process requirements (e.g., recycle to a pre-anoxic tank for denitrification will usually be at the higher end of this range).
- If the WAS pumps are equipped with speed control (i.e., VFDs), the operator should strive to operate as close to the BEP as possible on a consistent basis.
  - ▲ In the case of dedicated WAS pumps, the function of the pumps is to ensure the mass of biosolids calculated by the process operator is removed every day. Most pump types, particularly centrifugal pumps used in wastewater facilities, have a BEP. The BEP is defined by the pump manufacturer for the specific model and configuration of pump supplied. Because the instantaneous rate of wasting is not usually critical, most plants will find the best energy efficiency for dedicated WAS pumps by operating at constant speed as close as possible to the best efficiency point.

To assess the optimization status of energy efficiency relative to the secondary clarification process, plant staff are encouraged to consider the following questions:

- Do the RAS pumps have speed control and is the operation tied to the raw wastewater influent flow rate?

- If the plant has dedicated WAS pumps, do they operate consistently as close to the BEP as possible?
- Has plant/utility staff evaluated potential upgrades for increased energy efficiency?

Action

If areas of the secondary clarifier operation are considered Partially Optimized or Not Optimized, the following steps are recommended for implementing an action plan.

- Specific performance targets should be developed for the secondary clarifier operation. At many CAS plants, the secondary clarifier is the final step in the treatment process prior to disinfection. Therefore, the effluent discharge limits will dictate the performance targets for this unit process. Some of these performance targets are operational objectives and the operators should first consult existing plant design documents for original design parameters. The following performance targets should be identified:
  - Clarifier effluent TSS concentration.
  - Maximum solids loading rate and surface overflow rate.
  - Biomass inventory (aeration tank plus clarifier) with a goal of minimizing the biomass in the clarifier as much as is practically possible.
  - Minimum clarifier solids residence time.
  - Minimum TSS concentration in the RAS.
- A sampling, testing, and monitoring program should be developed for the plant that includes critical secondary clarifier data requirements. At minimum, the following data should be collected on a regular basis (daily is recommended) to monitor the process operation relative to the stated performance targets:
  - Clarifier core concentration, collected using a core taker.
  - Clarifier influent flow rate.
  - Clarifier influent TSS concentration.
  - Clarifier effluent TSS concentration.
  - Settled sludge level.
  - Settled sludge (or RAS) TSS concentration.
- The operators should ideally perform daily settling tests on samples of the mixed liquor entering the secondary clarifier.
  - The settling test performance should be logged and tracked over time to help determine changes in the biology that may be leading

  to settling issues. Poor settling sludge will directly and negatively impact the effluent being discharged to the environment.
  - If sludge settling declines, steps should be taken to identify any problematic (i.e., filamentous) bacterial strains in the mixed liquor population. Microscope observation or the assistance of an external laboratory will help to identify problematic bacteria.
- The data should be collected and entered into the appropriate spreadsheet on a daily basis. The data should be displayed on trend charts along with the respective performance targets to track the effectiveness of the process control and optimization programs.
- Operation of the RAS system should be optimized to meet the objectives of the total sludge mass control concept. Recycle rates should be tracked over time along with data for sludge blanket depth in the clarifier, the RAS TSS concentration, and the distribution of biomass between the clarifier and the aeration tank. With adequate data, the operator will be able to visualize how the RAS rate impacts the overall performance and the solids management in the system. Some general guidelines are as follows:
  - If the biomass in the secondary clarifier is increasing, a series of special studies should be completed. In the first study, reduce the RAS flow rate (by approximately 10 percent increments) to see if the RAS concentration can be increased, resulting in a reduced amount of biomass in the clarifier (i.e., rat holing due to excessive RAS flow rate was thinning the RAS concentration and causing the sludge mass in the clarifier to increase).
  - If the first study does not reduce the sludge mass in the clarifier, a second follow-up study is recommended. In the second study, the RAS flow rate is increased (by approximately 10 percent increments) to see if the RAS concentration can be maintained at the higher RAS flow rate, resulting in a reduced amount of biomass in the clarifier.
  - For both studies, the clarifier solids residence time is an important parameter used to evaluate the results.
  - In general, the RAS TSS should be at least two to three times the MLSS concentration in the aeration tank for a well-settling sludge. If there is little increase in RAS concentration and the sludge mass is very low in the clarifier, the RAS rate may need to be decreased.
- Operators are encouraged to use the total sludge mass control concept on a daily basis. The approach is more comprehensive and will result in a

higher level of expertise and process control. (Spreadsheet tools are available to assist with the data collection and calculations.)

- A routine maintenance and inspection program should be developed for the facility that includes the secondary clarifier. Inspection items should include monitoring of the scum buildup, checking the effluent weirs for buildup or biofilm, and observing the operation of mechanical systems, such as scum skimmers.

SOPs for the secondary clarifier should be developed, including an emergency response plan. For the secondary clarifier, it is specifically recommended that the emergency response plan include procedures for dealing with poor-settling sludge, which can have a significant impact on discharged effluent quality and, potentially, compliance. The operator along with plant management are encouraged to take an active role in developing SOPs. The operator should regularly review the procedures and the emergency response plan to be better prepared in the event of an unexpected issue with the system. The operations team is encouraged to continually update and improve SOPs and response plans. Well-documented SOPs will provide better continuity when staff changes occur within the facility.

Performance Limiting Factors

Table 4-4 summarizes factors related to the secondary wastewater treatment process that may be limiting performance or energy efficiency of the plant's secondary treatment processes. Check whether these factors are Optimized and Documented, Partially Optimized, or Not Optimized. Factors identified as Partially Optimized or Not Optimized will be prioritized in chapter 8, Identification and Prioritization of Performance Limiting Factors/Action Plan Development, where an action plan can be developed and implemented that allows for optimization of these parameters.

Nutrient Removal

Nutrient removal is an additional operational concern for many plants with sensitive receiving environments, and it is becoming more prevalent within the industry as environmental impacts are better understood and quantified. Nutrient removal for wastewater facilities refers specifically to nitrogen and phosphorus, and the following sections describe the most common treatment methods for their removal. A wide variety of process configurations exist to accomplish nutrient removal at WWTPs. A detailed discussion and comparison of these processes is beyond the scope of this guide. Readers are encouraged to consult the literature for additional information pertaining to wastewater treatment processes to achieve nutrient removal. The information and self-assessment questions provided in this guide are generally applicable to a variety of nutrient removal

Table 4-4. Performance limiting factor summary: secondary treatment

Self-assessment category	Questions for gauging optimization status	Optimization status		
		Optimized and Documented	Partially Optimized	Not Optimized
Suspended growth (CAS)	Have performance targets been developed for the secondary suspended growth treatment process?			
	Does the plant's monitoring and testing program allow for sufficient data collection to thoroughly monitor actual plant performance compared with the defined performance targets?			
	Are the appropriate trend charts being developed and populated to capture all of the raw data and calculated performance parameters to properly monitor the aeration tank?			
	Is the plant equipped with an appropriate chemical dosing system or other means to combat foaming or filamentous events?			
	Has a regular maintenance and inspection program been developed for the plant that includes daily inspection of the aeration tank and associated equipment?			
	Have SOPs been developed for the aeration tank, including an emergency response plan?			
Suspended growth (SBR specific)	Have the appropriate cycle and phase durations been identified and optimized for actual plant operation?			
	If the plant has a nonfixed style decanter (i.e., floating or adjustable), has the operation been optimized?			
	Has the batch volume per cycle been optimized to suit the actual operation and number of cycles per tank?			
Attached growth	Have performance targets been defined for the process?			
	Has a sampling and testing program been developed for the attached growth process that allows actual performance to be monitored compared to the defined targets?			

Self-assessment category	Questions for gauging optimization status	Optimization status		
		Optimized and Documented	Partially Optimized	Not Optimized
Attached growth *(continued)*	Has the collected data been plotted on trend charts along with any applicable performance targets?			
	Does the plant have the capability to provide sufficient aeration to meet the treatment requirements?			
	Does the attached growth process require regular replacement of the media and does the plant have sufficient spare media on site?			
	Has a regular inspection and maintenance program been developed for the attached growth process?			
	Have SOPs been developed for the attached growth process, including an emergency response plan?			
Secondary clarification	Have performance targets been developed for the secondary clarifier?			
	Has a sampling, testing, and monitoring program been put into place for the secondary clarifier?			
	Has a sampling and monitoring program been developed to monitor the settling characteristics of the sludge?			
	Have the appropriate trend charts been developed to properly track the collected data and monitor the clarifier performance relative to the defined operating targets?			
	Has the operation of the RAS system been developed to optimize the distribution of biomass between the aeration tank(s) and the secondary clarifier, and to satisfy the goals of the total sludge mass control concept?			
	Has the sludge wasting program been developed on the basis of the total sludge mass control concept that includes operating at a target SRT?			

Continued

Self-assessment category	Questions for gauging optimization status	Optimization status		
		Optimized and Documented	Partially Optimized	Not Optimized
Secondary clarification *(continued)*	Have steps been taken to minimize filamentous bacterial growth that leads to poor settling?			
	Does the operator have the ability to dose chlorine for control of filamentous bacterial growth?			
	Has a maintenance and inspection program been developed for the secondary clarifier?			
	Have SOPs been developed and documented for the secondary clarifiers, including an emergency response plan?			
Process energy considerations: Secondary treatment	Is the current aeration diffuser system providing acceptable oxygen transfer efficiency?			
	Does the plant use dissolved oxygen monitoring for blower speed control?			
	Is the current aeration blower technology providing the best efficiency?			
	Do the RAS pumps have speed control and is the operation tied to the raw wastewater influent flow rate?			
	If the plant has dedicated WAS pumps, do they operate consistently as close to the BEP as possible?			
	Has plant/utility staff evaluated potential upgrades for increased energy efficiency?			

processes. Plants using a nutrient removal process should complete the portions of this section applicable to their facility. Although this section is aimed at plants with an objective of nutrient removal, those with an objective of nutrient recovery may also derive benefit from this section, with modification of the self-assessment questions to address plant-specific goals.

Biological Nitrification

Understanding

Biological nitrification refers to the conversion of influent ammonia-nitrogen (NH_3–N) to nitrate-nitrogen (NO_3^-–N). Ammonia is of particular concern in receiving waters as it can be toxic to aquatic life under certain concentrations and environmental conditions. When ammonia is dissolved in wastewater, it disassociates into two forms. The forms are known as ionized ammonia, or ammonium ion (NH_4^+), and un-ionized ammonia (NH_3). The un-ionized fraction of the total ammonia is the form that can be toxic to aquatic life at levels above 0.019 mg/L[1]. The ratio of ionized to un-ionized ammonia depends on the pH and temperature conditions in the water. At neutral pH and wastewater temperatures in the 0°C–25°C range, the vast majority of the ammonia is of the ionized form and is not toxic. Depending on the influent concentration of total ammonia, the environmental conditions, and mixing zone considerations in the receiving water, ammonia removal may be required. The requirement to remove ammonia is defined by the regulatory agency in the treated wastewater effluent discharge permit.

When ammonia is a concern, the treatment unit processes are designed to achieve biological nitrification. Nitrification converts the influent ammonia-nitrogen to a different nontoxic form known as nitrate-nitrogen. When WWTPs have a total nitrogen discharge permit requirement, nitrate-nitrogen removal is achieved in a subsequent denitrification step (described in more detail later in this chapter). The process of converting ammonia to nitrate is typically accomplished in the aeration tank where the conversion occurs simultaneously with BOD5 oxidation. The nitrification process is a two-step conversion. In step one, specific bacteria, typically *Nitrosomonas*, convert ammonia first to nitrite. In step two, other bacteria, typically *Nitrobacter*, convert the nitrite to nitrate. The equations for these microbially mediated chemical reactions follow. (Note that the nitrification process produces hydrogen ion as a product of the conversion from ammonia to nitrite, meaning that nitrification can consume alkalinity and potentially impact pH, a concept that is discussed later in this section.)

$$2NH_4^+ + 3O_2 \rightarrow 2NO_2^- + 2H_2O + 4H^+ \quad \text{(Step 1)}$$

$$2NO_2^- + O_2 \rightarrow 2NO_3^- \quad \text{(Step 2)}$$

The process requires a very high level of process stability and application of the total sludge mass control concept, particularly in low temperature conditions (such as during winter months in cold climates). The quantity of sludge mass required to support nitrifying bacteria in the aeration tank is calculated. The nitrifying bacteria are able to grow and thrive along with the heterotrophic BOD-oxidizing bacteria. The nitrifying bacteria are particularly sensitive to low temperature conditions and grow slowly relative to the BOD-oxidizing bacteria. For that reason, a sufficient mass of the nitrifying bacteria must be maintained in the aeration basin at all times to fully convert all of the influent ammonia to nitrate. The slow growth of these organisms means that, if a significant quantity

1 Canadian Water Quality Guidelines for the Protection of Aquatic Life – AMMONIA, Canadian Council of Ministers of the Environment (CCME), 2010. The guideline is accessible at the following link: http://ceqg-rcqe.ccme.ca/download/en/141

of biomass is lost from the system (e.g., excessive sludge wasting or solids washout in the effluent from the secondary clarifier), it can take a long time to regenerate sufficient nitrifying biomass required for ammonia conversion. If the biomass is lost in cold weather conditions, it may be impossible to regenerate until warm temperatures return. At low temperatures, the growth rate of nitrifying bacteria is suppressed. These problems can result in extended time periods during which the influent ammonia is not fully converted and excess ammonia is discharged from the facility. Therefore, it is even more critical in systems designed to nitrify that operation of the aeration tanks and secondary clarifiers be optimized using the total sludge mass control concept where a target sludge is maintained at all times. Application of these concepts will help the operator ensure sufficient nitrifying biomass is maintained in the system and as much of this nitrifying biomass is maintained in the aeration tank as possible, and avoid accumulating biomass in the secondary clarifier where no nitrification occurs.

Status

Biological nitrification can be a challenging process, requiring careful attention from the operator and consistent process control, particularly in cold weather operation. The nitrification process occurs within the same aeration tank where BOD removal occurs. Therefore, nitrification can be essentially regarded as an extension of the CAS process previously described in this chapter. All of the key operating concepts, including OLR and F:M ratio, apply to ensure the BOD removal process still functions properly. The application of solids retention time and total sludge mass control also apply to nitrification. A key difference is that, if the system is required to nitrify, it will require a longer operating SRT and often a higher biomass inventory compared with systems designed for BOD removal only. If the minimum biomass inventory can be sufficiently maintained, and the system can operate consistently at the longer target SRT for nitrification, the system should be able to consistently achieve the effluent ammonia limits, provided no design limitations are identified as part of the unit process capacity assessment. The nitrification reactions tend to proceed more slowly than the BOD-oxidation reactions, therefore sufficient HRT needs to be provided in the aeration tank for the reactions to proceed. If the aeration tank is found to be undersized for the flow rates being handled, the system may not be capable of achieving full nitrification, due to the insufficient reaction time available.

Application of the total sludge mass control concept becomes even more important for biological nitrification due to the sensitivity of the process. It is highly recommended that the data required for application of the sludge mass control program be collected and reviewed on a daily basis. Consistently applying the concepts is the single most important task for the operator in terms of long-term sustained nitrification.

In addition to the total sludge mass control concept, a number of other operational parameters require careful monitoring and control to ensure complete nitrification is achieved. Similar to the BOD-oxidizing bacteria, the nitrifying biomass requires a consistent supply of oxygen. Careful monitoring of the aeration system and mixing systems is required to maintain a healthy biomass. Many plants use dissolved oxygen sensors to continuously monitor the oxygen levels in the aeration tank. It is common to maintain a dissolved oxygen residual in the aeration tank of 2.0 mg/L. This will ensure sufficient oxygen is being delivered to the biomass. Excess DO measured in the aeration basin is an indication that the process oxygen demands for both the BOD-oxidizing bacteria and the nitrifying bacteria have been met.

An additional concern for a plant designed for biological nitrification is that alkalinity is consumed through the ammonia conversion process. A certain level of alkalinity is required in the treatment process for pH buffering to maintain the pH in the recommended range of 6.8–7.5. Operating at pH levels outside of this range can be toxic to both the BOD-oxidizing and -nitrifying bacteria. In a CAS system where only BOD oxidation is required, the amount of background alkalinity in the raw wastewater is usually sufficient to maintain the neutral pH required. However, the consumption of alkalinity in a nitrification process can often lead to an alkalinity deficiency in the plant that will manifest as reduced pH

Case Study: Maintaining Alkalinity to Support Nitrification

Nitrification occurs in the same aeration tank where BOD removal occurs. Essentially, nitrification can be regarded as an extension of the CAS process previously described in this chapter. The application of solids retention time and the total sludge mass control apply to nitrification. The key difference is that if the system is required to nitrify, it will require a longer operating SRT compared with systems designed for BOD removal only. If the minimum biomass inventory can be sufficiently maintained, and the system can operate consistently at the longer target SRT required for nitrification, the system should be able to consistently achieve the effluent ammonia limits, provided no design limitations are identified as part of the capacity assessment. The nitrification reactions tend to proceed more slowly than the BOD-oxidation reactions, therefore sufficient HRT needs to be provided in the aeration tank for the reactions to proceed. If the aeration tank is found to be undersized for the flow rates being handled, the system may not be capable of achieving full nitrification due to the insufficient reaction time available.

An additional complication for a WWTP that is required to achieve biological nitrification is that alkalinity is consumed through the ammonia conversion process. Every mg/L of TKN removed via nitrification consumes 7.14 mg/L of alkalinity. The consumption of alkalinity in a nitrification process may lead to an alkalinity deficiency in the plant that will manifest as reduced pH in the

Continued

aeration tank. If there is an alkalinity deficiency, alkalinity must be supplemented to the process in the form of chemical addition. A number of chemicals are commercially available for alkalinity supplementation, including sodium hydroxide, soda ash, and sodium bicarbonate. Figure 4-11 displays a graph of the raw wastewater alkalinity compared with the minimum alkalinity required to support nitrification for the case study facility. The more highly variable line on the graph represents the daily raw wastewater alkalinity concentration, and the horizontal trace is the theoretical minimum alkalinity required to support nitrification, based on an average raw wastewater TKN concentration of 32.5 mg/L. The trend chart confirms that on multiple days the actual raw wastewater alkalinity concentration was less than the theoretical minimum alkalinity required to support nitrification.

It is advisable for utility staff to calculate the minimum alkalinity concentration required to support nitrification, based on the raw wastewater TKN. For example, based on an average raw wastewater TKN concentration of 32.5 mg/L, approximately 232 mg/L of alkalinity is required to support nitrification (i.e, 32.5 mg/L TKN * 7.14. mg/L alkalinity = 232 mg/L). If the actual alkalinity is more than 10 percent higher than the theoretical concentration required, then there is adequate alkalinity to achieve nitrification. If it is determined that the concentration of alkalinity is marginal, utility staff should measure the alkalinity concentration in the secondary effluent. Metcalf and Eddy specifies a concentration of 80 mg/L of alkalinity in the secondary effluent (Tchobanoglous et al. 2014). If the measured secondary effluent alkalinity is less than 80 mg/L, utility staff should investigate options for adding supplementary alkalinity upstream of the aeration basin.

All of the above status items are well within the operators' control and should be monitored consistently on a daily basis. An additional, and often overlooked, concern for the nitrification process is the presence of inhibitory substances in the raw wastewater, as previously discussed in this chapter. These substances are of particular concern because they can be challenging to detect and identify as the root cause of the problem. It is common, when there are issues with nitrification, to assume that something at the plant must be the cause of the problem (e.g., cold weather, insufficient biomass, solids washout, etc.). However, all of the parameters within the operators' control may be fully optimized and the nitrification process may still perform poorly as a result of nitrification inhibitory substances being discharged by users of the collection system. The best way to detect this problem is through diligent data collection and a very thorough and consistent process control program. By applying the concepts in this guide, the operators will be able to demonstrate that consistent operation of the nitrification process is being achieved and better identify if performance problems are due to outside influences.

A benefit to utility staff of establishing consistent and stable process control by implementing the total sludge mass control concept is that cause-and-effect relationships are more clearly detectable. A city WWTP evidenced this when wastewater production at an industry in the city resulted in a reduction in the nitrification efficiency at the treatment plant. The plant uses four liquid trains, and the total sludge mass control concept was implemented in all four trains to ensure the level of process control was adequate to achieve complete nitrification. The typical influent ammonia concentration is approximately 25 mg/L. Normally, the facility is able to achieve stable and complete nitrification with typical treated effluent ammonia concentrations of less than 1 mg/L as illustrated in Figure 4-12. Starting in late October and early November 2009, there was a significant negative impact on the nitrification efficiency, resulting in combined treated effluent ammonia values greater than 20 mg/L. This implies a virtual shutdown of nitrification in the four liquid trains because the influent and final effluent ammonia concentrations were virtually identical.

Further investigation by the operations staff confirmed that the ammonia concentration had increased abruptly and simultaneously in all four liquid trains. This is illustrated in Figure 4-13 by stacking the secondary effluent ammonia concentration for all four liquid trains on the same time line as Figure 4-12. Prior to October 2009, the rate of nitrification was high and very stable. This is evidenced by the virtual absence of measurable ammonia in all four secondary effluents in Figure 4-13. This pattern was identical to the combined final effluent ammonia concentration depicted in Figure 4-12.

The impact of the loss of nitrification was serious and resulted in the plant exceeding the monthly average final effluent discharge standards for October 2009. The abrupt increase was untenable and it prompted a full-scale investigation for potential outside sources capable of having such a negative impact on the nitrification process. Utility staff were justified in investigating outside sources, because the level of process control at the WWTP was unchanged prior to, and following, the abrupt increase in ammonia concentration. More importantly, the increase in ammonia concentration had occurred simultaneously in all four liquid trains, confirming that the cause of the loss of nitrification most likely originated outside the WWTP. While the events in late October 2009 were the most serious in terms of not achieving compliance with the discharge standards, Figures 4-12 and 4-13 also confirm that several more minor events occurred in December 2009 that caused the ammonia concentration to increase in the secondary and combined final effluent.

Subsequent studies confirmed that the phenomenon causing the loss in nitrification was actually due to inhibition of the nitrifying bacteria. The source of the inhibitory material was an industry in the city. At subsequent meetings with industry officials, utility staff used the same trend charts to clearly illustrate the cause and effect relationship they had detected at the WWTP, and to justify why they could confidently conclude that the source of the nitrification inhibition originated outside of the treatment plant.

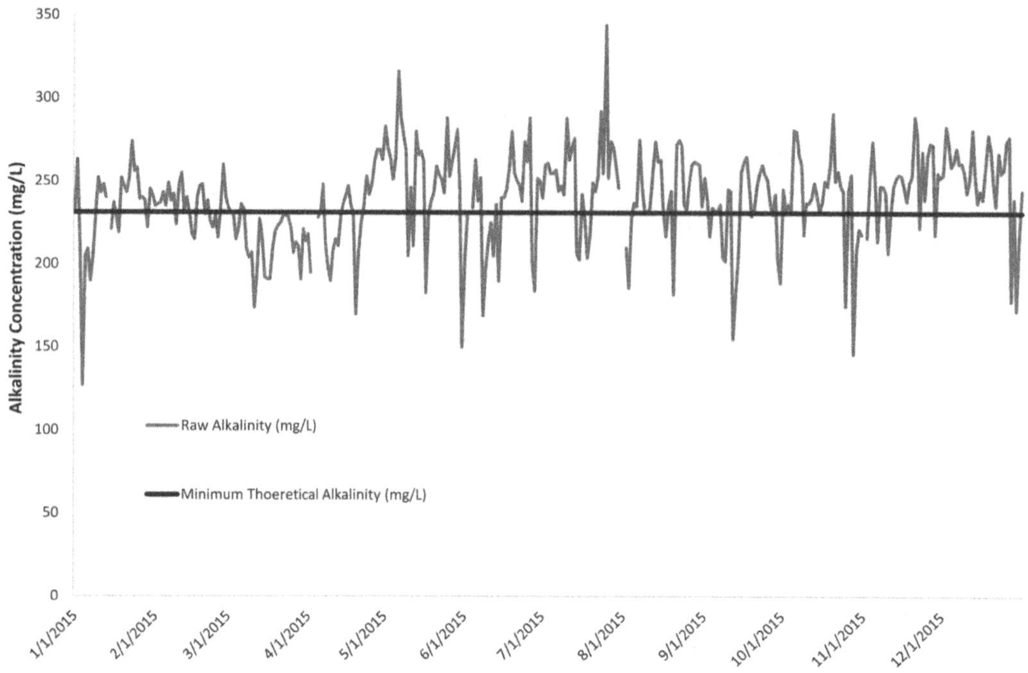

(Source: Gerard Wheeler, Global Faciliation, Inc.)

Figure 4-11. Raw wastewater alkalinity concentration compared with theoretical minimum concentration required to support nitrification

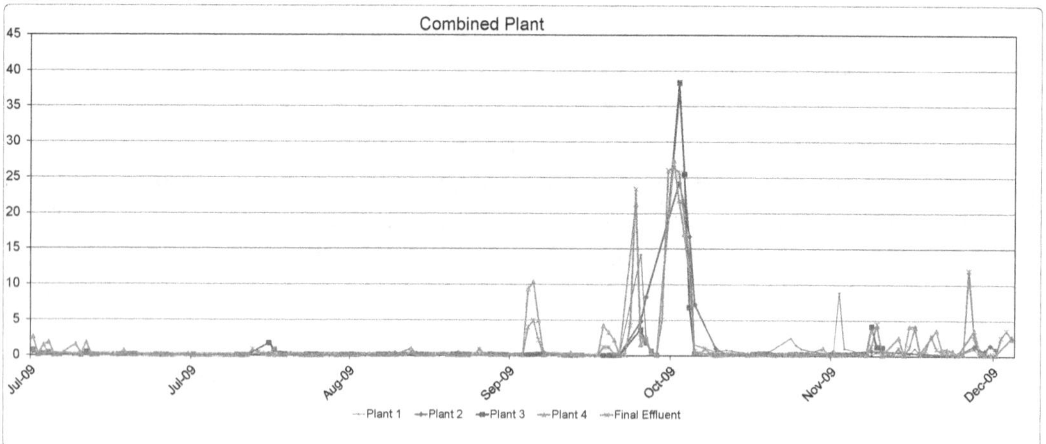

(Source: Gerard Wheeler, Global Facilitation, Inc.)

Figure 4-12. WWTP combined final effluent ammonia concentration, July to December 2009

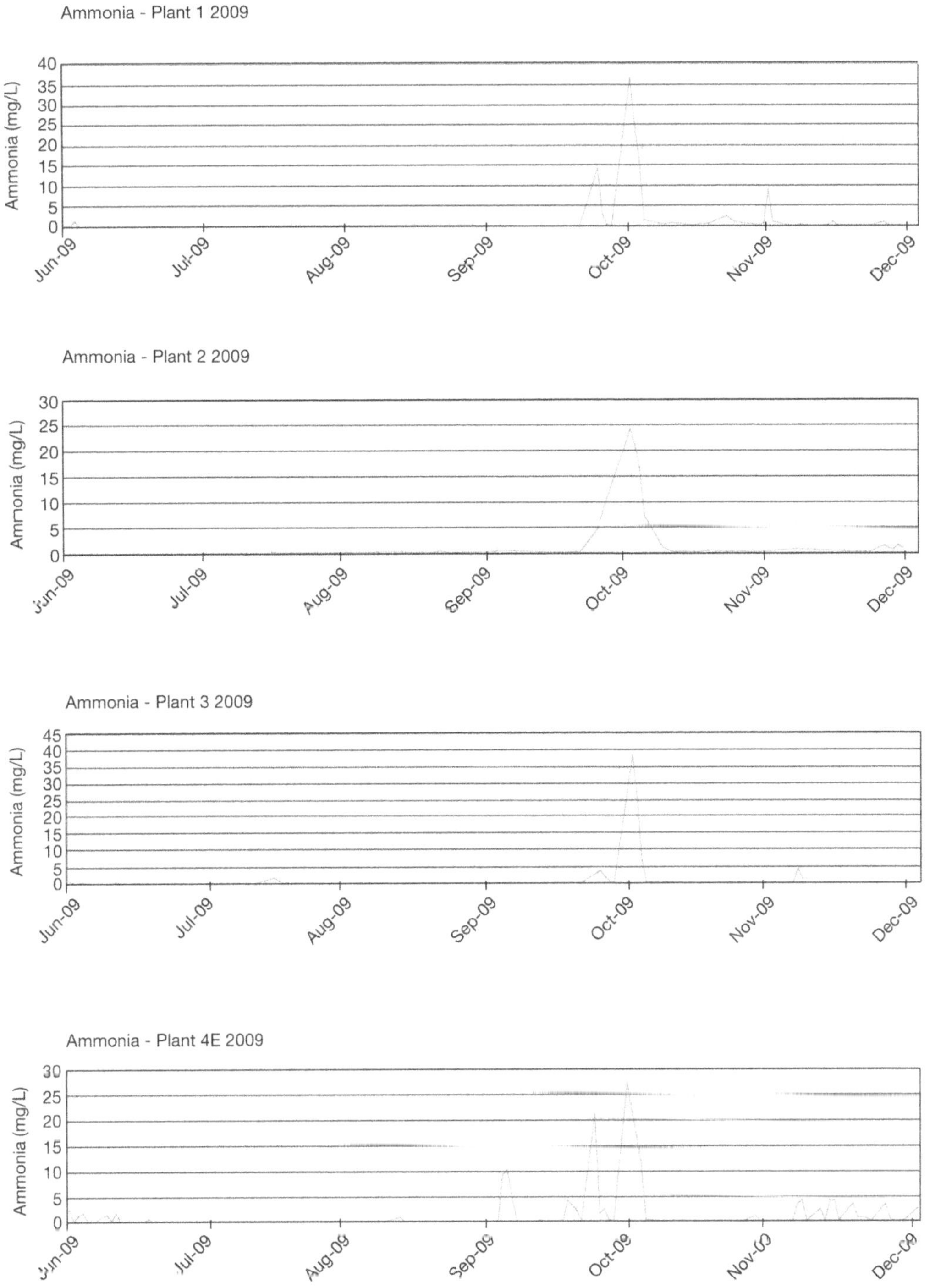

(Source: Gerard Wheeler, Global Facilitation, Inc.)

Figure 4-13. WWTP liquid trains 1–4 secondary effluent ammonia concentration, July to December 2009

Total Sludge Mass Control Concept Demonstrated Due Diligence

At the subsequent meeting to review the noncompliance event with local regulatory enforcement staff, the same total sludge mass control spreadsheet and trend charts were used to explain the event. The materials helped communicate how the process control and data collection procedures enabled utility staff to quickly detect that the nitrification process in all four liquid trains were impacted at the same time and to the same degree.

More importantly for the enforcement staff, the data also showed that complete nitrification was being achieved immediately before and after the event. This convinced enforcement staff that utility staff had an adequate barrier of activated sludge in place to achieve complete nitrification. After inspecting the process control spreadsheet and trend charts, the enforcement staff concluded that utility staff had demonstrated due diligence, and they decided not to proceed with legal enforcement of the noncompliance event.

in the aeration tank. In this case, alkalinity must be supplemented to the process in the form of chemical addition. A number of commercially available chemicals may be used for alkalinity supplementation, including sodium hydroxide, soda ash, and sodium bicarbonate. The alkalinity requirement case study provides one example of how this factor was addressed by one wastewater treatment facility.

The following questions are designed to help the operator better understand the optimization status of the nitrification process. Operators are encouraged to develop and consider additional questions that may be specific to their utility:

- Have performance targets been developed for the nitrification process? Key performance targets are recommended to include the following:
  - Effluent ammonia-nitrogen (NH_3–N) concentration.
  - Target SRT and biomass inventory for total sludge mass (aeration tank + clarifier).
  - Aeration Tank DO Concentration.
  - Aeration tank hydraulic retention time.
- Has a sampling, testing, and monitoring program been put into place that includes the required data for nitrification? The program should include monitoring of parameters required for the total sludge mass control concept introduced in this guide. A summary of the required parameters is as follows. (Note that some of these parameters overlap with the list provided for the secondary treatment suspended growth and secondary clarification subsections.)
  - Influent and effluent NH_3–N concentrations.
  - Aeration tank MLSS and MLVSS concentrations.

  - Aeration tank DO concentration.
  - Clarifier settled sludge/RAS MLSS concentration and sludge blanket depth.
  - Influent and effluent alkalinity concentration (expressed in mg/L as $CaCO_3$).
- Have the appropriate trend charts been developed to collect the required data and track actual system performance relative to the defined operating targets?
- Has a maintenance and inspection log been developed for the plant that includes the secondary treatment system and nitrification process?
- Have SOPs been developed for the nitrification process, including an emergency response plan?

Process Energy Considerations

The biological nitrification process is an extension of the activated sludge process whereby key operating parameters such as SRT are adjusted to encourage the growth of a nitrifying biomass. The major process energy considerations, therefore, are essentially the same as those identified in the suspended growth secondary treatment section of this guide. The largest energy consumer will be the aeration system, and any optimization of the aeration control and blower speed could provide significant benefits in terms of energy efficiency and associated costs.

Action

If areas of the nitrification process are considered Partially Optimized or Not Optimized, the following steps are recommended for developing and implementing an action plan.

- Performance targets for the process should be developed and clearly defined, with a margin of error with respect to permit requirements. The main objective of the nitrification process is the removal of ammonia, and this is often defined by the operating permit for the facility. Additional operating targets may be defined in original design documents for the facility and operators should consult these records first. The following performance targets should be defined:
  - Effluent NH_3–N limit: It is common to require a concentration of <1.0 mg/L NH_3–N.
  - Target SRT: Typically in the range 10 to 15 days for nitrification.

  - Target biomass inventory: A unique requirement for each plant as defined through application of the total sludge mass control program.
  - Aeration tank DO Concentration: Typically 2.0 mg/L but could target 0.5 mg/L or 1.0 mg/L for energy savings provided there is a reliable and high level of process monitoring and control.
- A sampling and testing program should be developed for the facility that allows for consistent (daily is recommend) collection of data for a high level of process monitoring and control. Ensuring the data are collected on a daily basis for application of the total sludge mass control concept and daily WAS volume calculation is most critical for successful nitrification. It is recommended that the following data be collected daily:
  - Influent and effluent NH_3–N concentrations.
  - Aeration tank MLSS and MLVSS concentrations.
  - Aeration tank DO concentration.
  - Clarifier settled sludge/RAS MLSS concentration and sludge blanket depth.
  - Influent and effluent alkalinity concentration (expressed in mg/L as $CaCO_3$).
- The collected data should be entered into a spreadsheet on a daily basis and trend charts should be updated and evaluated. The WAS volume should be calculated daily using the solids mass control spreadsheet tool available to download by PCW subscribers for use with this guide. All performance targets for the process should be identified on trend charts along with the performance target values to provide a quick visual indication of actual performance. Due to the significant challenge of maintaining consistent nitrification, extra care and attention should be paid to the trend charts to observe any changes in the operation, no matter how small they may seem.
- A regular maintenance and inspection program should be developed for the facility, which includes the aeration tank where nitrification occurs. The inspection items have been previously defined in the secondary treatment suspended growth subsection of this guide.
- SOPs for the nitrification process should be developed, including an emergency response plan. It is recommended, for nitrification specifically, that the emergency response plan include procedures for dealing with increased effluent ammonia levels, particularly in cold weather if applicable. The operator is encouraged to take an active role along with plant management in developing SOPs. The operator should regularly review the procedures and the emergency response plan to be better prepared in the event of an unexpected issue with the system. The

operations team is encouraged to continually update and improve SOPs and response plans and ensure that training is provided to all pertinent members of plant staff. Well-documented SOPs will provide better continuity when staff changes occur within the facility.

Biological Denitrification

Understanding

Depending on the sensitivity of the receiving water, some WWTPs may be required to achieve a higher level of treatment. In addition to nitrification, an additional treatment step, called denitrification, may be required. The combination of nitrification and denitrification achieve complete nitrogen removal. The process of nitrification does not actually remove nitrogen from the wastewater but converts it from ammonia to nitrate. This process protects aquatic life in the receiving water from potentially toxic levels of un-ionized ammonia. The generation of excess nitrates (a byproduct of nitrification) can also lead to detrimental impacts on some receiving waters depending on the classification of the receiving water body (e.g., drinking water source, high recreational use, etc.). Ingestion of water with excess levels of nitrate can cause serious health issues, particularly in young children. Nitrate is also a key nutrient leading to eutrophication of lakes and rivers. It is becoming more common for WWTPs that nitrify to be also required to achieve denitrification for nitrate removal. The general chemical reaction for denitrification, similarly a microbially mediated reaction, is as follows:

$$6NO_3^- + 5CH_3OH \rightarrow 3N_2\,(g) + 6OH^- + 5CO_2 + 7H_2O$$

Nitrate can be removed from the effluent discharge through denitrification using anoxic (i.e., low oxygen) conditions. The absence of oxygen promotes the growth of denitrifying bacteria that convert nitrate to nitrogen gas, which is subsequently released to the atmosphere. The denitrification process can take place either before the aeration tank (commonly referred to as a pre-anoxic process) or after the aeration tank (commonly referred to as a post-anoxic process). If pre-anoxic denitrification is used, a recycle loop must be used to return nitrate-rich mixed liquor from the aeration tank back to the pre-anoxic tank. The amount of nitrate removed is dependent on the recycle flow rate compared with the effluent flow rate. The recycle flow rate must be high enough to return the required mass of nitrate to the pre-anoxic tank. There are practical limitations to how much recycle flow can be returned. In some cases, depending on the level of nitrate removal required, the existing RAS flow rate may be sufficient to return the mass of nitrate required in the pre-anoxic tank. This configuration is referred to as a single-loop pre-anoxic denitrification system. In other cases, the existing RAS flow rate is not high enough to return the mass of nitrate required in the pre-anoxic tank. For this condition a second RAS recycle loop is often added. This configuration is referred to as a double-loop pre-anoxic denitrification system. Simplified flow diagrams of these configurations are provided in Figures 4-14 and 4-15.

If it is not practical to remove all of the nitrate required in the pre-anoxic zone, post-anoxic denitrification may be required as an additional polishing step. In this configuration, the remaining nitrate that is not removed in the pre-anoxic zone is removed in a second anoxic tank downstream of the aeration tank(s). Post-anoxic denitrification normally requires the addition of a supplemental carbon source to provide food for the anoxic bacteria. When denitrification occurs in the pre-anoxic tank, the carbon required is provided by the raw wastewater BOD_5. Since post-anoxic denitrification occurs after the aeration tank, virtually all of the soluble BOD_5 has already been removed. Therefore, no carbon source remains to feed the post-anoxic bacteria. Supplemental carbon in the form of chemical addition is required. Most commonly, methanol is used as the carbon source for post-anoxic denitrification, as displayed in the previous denitrification chemical equation. The hazardous nature of handling methanol has led to the development of some proprietary alternative chemicals, such as MicroC™. Corn syrup can also be used as a source of supplemental carbon, although it is not common within the industry.

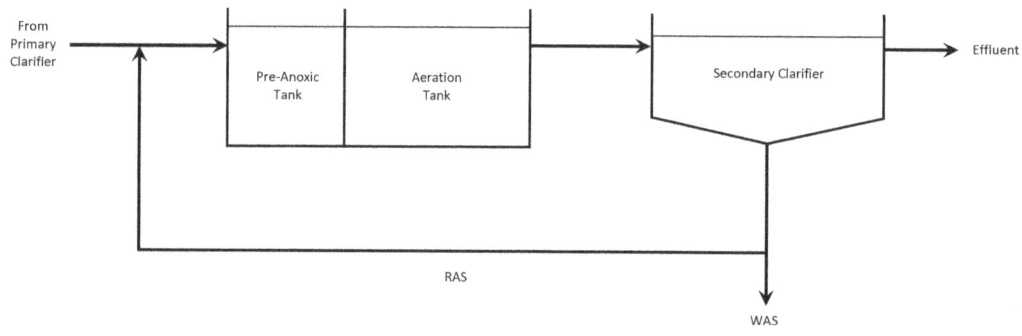

Figure 4-14. Single-loop pre-anoxic denitrification system flow diagram

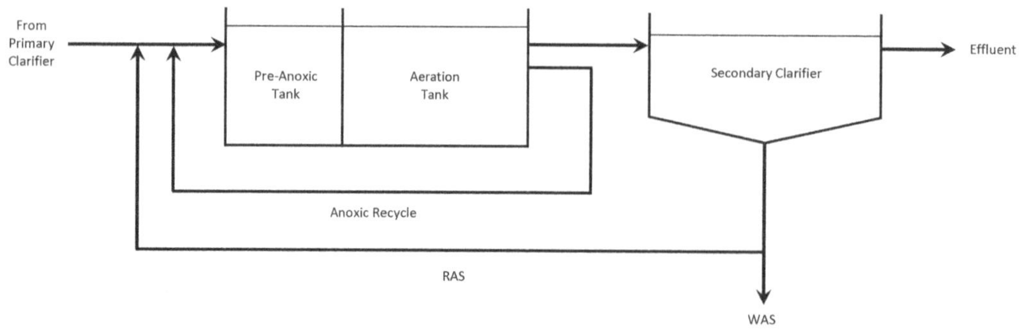

(Source: Daryl Burke, EXP Services, Inc.)

Figure 4-15. Double-loop pre-anoxic denitrification system flow diagram

Process control of the denitrification process includes calculating the target quantity of biomass necessary to achieve the required level of nitrate removal in the anoxic tank. Maintaining zero or a very low dissolved oxygen condition in the anoxic tank is critical to maximizing the effectiveness of the denitrification process. An essential aspect of process control is effective mixing of the anoxic tank to ensure the denitrifying bacteria are in suspension and in contact with the nitrate in the anoxic tank.

Achieving denitrification introduces an additional complication with respect to alkalinity. While nitrification is a net consumer of alkalinity, the denitrification process actually produces alkalinity. As a general rule, if complete nitrification and denitrification are achieved, the denitrification process will produce approximately half the amount of alkalinity that was consumed by nitrification. Even with the production of alkalinity by denitrification, the overall result may still result in a net deficit of alkalinity that needs to be supplemented. The production of alkalinity via denitrification can help reduce the amount of chemical required to supplement alkalinity. In some cases, if enough alkalinity is produced, the need for chemical supplementation can be eliminated.

An additional benefit of the denitrification process, when implemented as a pre-anoxic configuration, is that it tends to discourage the growth of filamentous bacteria in the activated sludge process. This can improve settling in the secondary clarifier and reduce the amount of chlorine required for filamentous bacteria control.

Status

The overall goal of biological nitrification and denitrification is to achieve complete nitrogen removal. Effective denitrification can only be achieved if the nitrification process is also fully optimized. An optimum level of nitrification can be achieved if a sufficient quantity of biomass is maintained in the aeration tank at the target Solids Residence Time (SRT). Maintaining a stable biomass will ensure the nitrification process is optimized and the concentration of nitrate is maximized. Once the nitrate production is maximized, the operations team can begin to focus on optimizing the denitrification process in order to convert nitrate produced during the nitrification to nitrogen gas.

Many WWTPs are capable of achieving up to 75 percent removal of total nitrate in the pre-anoxic zone. If post-anoxic denitrification is required, it is generally included as a polishing step to remove nitrate down to very low levels. Therefore, optimizing the pre-anoxic zone to maximize nitrate removal should be the first priority of the operations team. It is desirable to rely as little as possible on post-anoxic treatment for nitrate removal because this will minimize the supplemental carbon dose required. The amount of nitrate removed in the pre-anoxic zone is a function of four things:

- The mass of nitrate being returned from the aeration process.
- The HRT in the pre-anoxic tank.

- The quantity of denitrifying biomass in the pre-anoxic tank.
- Climatic conditions.

The amount of nitrate returned from the aeration tank is dependent on the recycle flow rate of activated sludge. In some plants, such as MBR facilities, the RAS flow may be great enough to return a sufficient mass of nitrate for denitrification. In most CAS systems where the RAS flows are relatively low, a separate mixed liquor recycle loop is required. Typically, the mixed liquor recycle loop is taken from the aeration tank and pumped back to the pre-anoxic tank, while at the same time the RAS system pumps nitrified sludge from the secondary clarifier back to the aeration tank. In either configuration, control of the recycle flow rate determines the mass of nitrate returned to the pre-anoxic tank for denitrification. Additionally, the pre-anoxic tank HRT must be greater than the reaction time required for the denitrification process to be completed. The influent wastewater flow rate, combined with the recycle sludge flow rate, inversely impact the HRT (i.e., as the flow of either or both increases, the HRT decreases). Therefore, a practical limit to the recycle sludge flow rate exists, because it cannot exceed the reaction time required for the denitrification process to be complete.

The quantity of biomass plays a crucial role in the denitrification process stability and performance. In the anoxic zone, the F/M ratio for the denitrifying biomass impacts the Specific Denitrification Rate (SDNR). The SDNR is defined as the mass of nitrate that can be removed per unit of biomass (i.e., g NO_3^-–N/g biomass•day). The higher the F/M ratio, the higher the SDNR, and the more nitrate that is removed in the process. In practice, the operator has little control over the mass of BOD (i.e., food being fed to the system). Therefore, the approach for managing the F/M ratio is to control the biomass in the tank. If the F/M ratio for the pre-anoxic tank is too low (i.e., less than 1 g BOD_5/g biomass), the SDNR may be less than optimal, resulting in lower removals of nitrate. In practice, maintaining the biomass at a scientifically derived target is a priority for the operator to ensure maximum removal of nitrate.

Where post-anoxic denitrification is required, a carbon source must be introduced to provide the necessary food for the denitrifying biomass. A post-anoxic system has no recycle loop, so the mass of nitrate entering the post-anoxic tank is the total nitrate produced in the aeration tank via nitrification minus the amount of nitrate recycled to the pre-anoxic tank. Post-anoxic denitrification also requires a short post-aeration step to encourage the release of nitrogen gas to the atmosphere. A key parameter the operator must control in post-anoxic denitrification is the supplemental carbon dose, while maintaining an adequate mass of denitrifying bacteria in suspension to remove the nitrate.

The following questions are designed to help the operator better understand the optimization status of the biological denitrification process. Operators are encouraged to develop and consider additional questions that may be specific to their utility:

- Have performance targets been developed for the biological denitrification system?
  - Typically, performance targets reflect the limit for final effluent nitrate concentration specified in the discharge permit, with a safety margin incorporated.
  - Some plants may choose to denitrify to generate alkalinity and offset chemical demand. In these cases, the original design documentation may assist with defining performance targets.
- Has a sampling and testing program been developed that includes the denitrification process? Sampling and testing should be completed on a daily basis and include the following parameters:
  - Nitrate concentration in mixed liquor leaving the aeration tank.
  - Final effluent nitrate concentration.
  - Nitrate concentration leaving pre-anoxic tank.
  - Mixed liquor TSS in pre-anoxic and post-anoxic tanks, if applicable.
- Have all data been plotted on trend charts along with applicable performance targets? Trend charts should be updated daily.
- Has the nitrification process been optimized in accordance with the sludge mass control concept?
- Has a nitrate mass balance been completed for the pre-anoxic denitrification process to ensure sufficient recycle of nitrate from the aeration tank?
- If the plant includes a post-anoxic denitrification step, has the pre-anoxic tank been optimized for maximum nitrate removal? Has the appropriate testing and data collection been implemented to verify nitrate removal in each anoxic zone?
- If post-anoxic denitrification is used, has the supplemental carbon dosing system been optimized for chemical usage efficiency?
- If methanol is used for supplemental carbon dosing, have alternative, less hazardous chemical options been explored? Alternative chemicals may offer benefits over methanol.
- Has a routine maintenance and inspection program been developed that includes the denitrification process? Inspection items should include the following:
  - Visual inspection of anoxic tanks for adequate and even mixing.
  - Inspection of chemical dosing equipment for leaks.
- Have SOPs been developed for the denitrification process, including an emergency response plan?

Action

If areas of the biological denitrification process operation are considered Partially Optimized or Not Optimized, the following steps are recommended for implementing an action plan:

- Performance targets should be clearly defined for the process, including required nitrate removals for both pre-anoxic and post-anoxic zones. Performance targets should also be clearly defined for the nitrification process, as this must first be optimized to achieve total nitrogen removal.
- A sampling and testing program should be implemented that allows for daily monitoring of key parameters for the biological denitrification process.
- All data should be plotted on trend charts and reviewed daily for any signs of reduced performance.
- In many plants, the denitrification process is the second step in a total nitrogen removal process. As a result, the preceding nitrification process must first be optimized before optimal total nitrogen removal can be achieved using both processes.
- A nitrate mass balance should be completed for the system as a way to verify that the appropriate mass of nitrate is being returned to the pre-anoxic tank. For plants that only use pre-anoxic denitrification, the mass rate of nitrate required to be recycled to the pre-anoxic tank equals the total mass rate of nitrate leaving the aeration tank minus the allowable mass rate of nitrate in the effluent (calculated as the effluent nitrate concentration limit times the effluent flow rate).
- In plants that use both pre-anoxic and post-anoxic denitrification, the optimal operation would see the nitrate removal maximized in the pre-anoxic zone. This will reduce supplemental carbon dosing requirements in the post-anoxic tank. Completing a nitrate mass balance will help with maximizing the mass of nitrate being recycled for removal in the pre-anoxic tank.
- The supplemental carbon dosing system for the post-anoxic tank should be optimized for efficient use of chemical while meeting the process demand. Metering of supplemental carbon chemical can be controlled via flow-pacing in proportion with the plant flow. As wastewater flow rate increases, so does the mass of nitrate generated in the aeration system. The increase in the mass of nitrate leads to an increased demand for carbon chemical in the post-anoxic tank. Online instrumentation, if available, and feedback from the various processes (e.g., nitrate probes) can be used to further optimize the flow rate of carbon chemical to the system. For example, when low influent flow rates and/or ammonia

concentrations exist, it may be possible for the pre-anoxic zone to accomplish 100 percent of the nitrate removal. This could potentially reduce the instantaneous demand for supplemental carbon chemical to almost zero.

- If the plant is using methanol as a source of supplemental carbon, there may be benefits for switching to an alternate chemical. Other chemical options can be less hazardous than methanol and possibly less expensive, due to less stringent transportation and storage requirements.
- A routine inspection and maintenance program should be developed for the plant to ensure mechanical issues are not contributing to reduced performance. Visual inspection of the process equipment should be completed daily.
- SOPs should be developed for the entire plant that includes the biological denitrification process. An emergency response plan should be developed that includes procedures for safe handling and storage of methanol (or other chemicals) as well as a response plan for chemical spills.

Biological Phosphorus Removal

Understanding

Similar to nitrogen removal, phosphorus in the raw wastewater is consumed by bacteria, under certain conditions, to promote new cell generation. Specific bacteria, referred to as phosphorus accumulating bacteria, are capable of storing excess quantities of phosphorus under these conditions. Biological phosphorus removal typically requires the addition of an anaerobic stage in addition to the CAS process. Wastewater from the plant headworks typically enters the anaerobic tank as the first stage in the process. The anaerobic step can then be followed either by an aeration tank or the combination of an anoxic-aerobic process if total nitrogen removal is also required. The biological phosphorus removal process is then followed by secondary clarification for solids removal and any tertiary treatment that may be required. A variation on the typical phosphorus removal configuration can include a sidestream anaerobic tank that is fed by the RAS line from the secondary clarifier.

The main purpose of the anaerobic step in the biological phosphorus removal process is to generate the preferred process conditions within the downstream aeration tank for the growth of phosphorus accumulating bacteria. In general, anaerobic fermentation results in the formation of specific volatile fatty acid byproducts that favor the development of the phosphorus accumulating bacteria. Once these favorable conditions are established, the phosphorus accumulating bacteria are able to generate new cells that further enhance the rate of phosphorus uptake in the wastewater.

Status

The biological phosphorus removal process uses an active biomass similar to CAS, and includes many of the same concepts of the total sludge mass control program. Consistently maintaining a target sludge mass in the bioreactor tanks and operating consistently at a target SRT are both critical process control parameters. Many systems designed for biological phosphorus removal are also designed for total nitrogen removal and include both nitrification and denitrification steps. The additional requirement of biologically removing phosphorus increases the importance of a higher and more scientific level of process control based on the principles of the total sludge mass control program. This is especially true given that a shorter SRT tends to favor biological phosphorus removal whereas biological nitrogen removal often requires a longer SRT. An effective balance must be established that meets the requirements of all aspects of the process.

One of the primary volatile fatty acids that is formed by the anaerobic fermentation step is acetate, which promotes the growth of the phosphorus accumulating bacteria. Acetate is produced in the anaerobic zone by breaking down biodegradable COD in the raw wastewater. During low organic loading periods, the production of acetate in the anaerobic zone can decrease to levels that are insufficient to sustain the bacterial population required for biological phosphorus removal. During these periods, acetate chemical is added to supplement the volatile fatty acid production in the anaerobic zone. More acetate production generally equates to more phosphorus accumulating bacteria generation and enhanced phosphorus removal.

Another influence on the performance of the biological phosphorus removal process is the concentration of nitrate in the RAS to the anaerobic reactor. Nitrate can reduce the volatile fatty acid production in the anaerobic reactor. This leads to less than optimal conditions for the growth of phosphorus accumulating bacteria. Nitrate is produced as a result of the nitrification process. It is recommended to avoid recycling nitrate-rich activated sludge from the secondary clarifier to the anaerobic zone. It is preferential to recycle sludge from the pre-anoxic zone if available where there should be minimal nitrate in the activated sludge.

There are practical limitations to the level of phosphorus removal that can be achieved from the biological phosphorus removal process. In cases where very low effluent total phosphorus concentrations are required, plants may be required to supplement the process with chemical addition for soluble phosphorus precipitation. Chemical phosphorus removal is discussed in more detail in the following section.

The process of biological phosphorus removal is the conversion of soluble phosphorus in the raw wastewater to a solid. These solids are bacterial cells that are then wasted as part of the overall sludge mass control program. Solids separation can be accomplished by a conventional clarifier or with more advanced technology such as tertiary filters or membranes. Regardless of what technology is used to remove solids, consistent and accurate removal of solids is critical for achieving the final effluent phosphorus target for the plant.

The following questions are designed to help the operator better understand the optimization status of the biological phosphorus removal process. Operators are encouraged to develop and consider additional questions that may be specific to their utility:

- Have performance targets been defined for the biological phosphorus removal process? Performance targets should define the total phosphorus discharge limit, and the level of removal that can be achieved biologically.
- Has a total sludge mass control program been implemented to define a target sludge mass and SRT that balances the requirements for biological phosphorus removal with total nitrogen removal? Operating at a reduced SRT may enhance biological phosphorus removal but conversely may limit total nitrogen removal.
- Has a sampling, testing, and monitoring program been developed for the biological phosphorus removal process? Sampling and testing is recommended to include the following parameters:
  - Final effluent total phosphorus;
  - Nitrate concentration in recycle stream to anaerobic zone;
  - Total phosphorus concentration in the aeration tank; and,
  - All parameters documented in the suspended growth treatment section for implementation of the total sludge mass control program.
- Has the plant optimized the acetate chemical dosing system to maximize phosphorus removal and maintain a consistent mass of phosphorus accumulating bacteria?
- Does the plant have the capability to supplement a chemical coagulant to enhance phosphorus removal? If so, has this system been optimized according to the chemical phosphorus removal section of this guide?
- Has the final effluent solids removal process been optimized in accordance with the applicable section(s) of this guide?
- Have SOPs been developed for the biological phosphorus removal process that include an emergency response plan? In particular, for the biological phosphorus removal process, the procedures and emergency response plan should include information on safe operating techniques for handling methane that is generated in the anaerobic fermentation process.

Action

If areas of the biological phosphorus removal process operation are considered Partially Optimized or Not Optimized, the following steps are recommended for consideration in developing and implementing an action plan.

- Performance targets should be defined for the system. In most cases, the plant's operating permit will define the total phosphorus limit that must be achieved in the final effluent. The plant operations staff should consult design documents for the facility to help determine the biological removal performance and any supplemental chemical phosphorus removal that may be required.
- Total sludge mass and SRT targets should be defined for the process that balance the requirements for biological phosphorus removal with nitrification and denitrification processes. A number of process variations have been implemented for biological nutrient removal (phosphorus + nitrogen). Operators should consult plant design documentation for design SRT values to use as an initial target for the plant.
- A sampling and testing program should be developed for the plant that includes critical parameters for the biological phosphorus removal process. The testing should be completed daily and should allow the operator to track the phosphorus removal as well as the full implementation of the sludge mass control program.
- If not currently in use, the plant should consider adding the capability to dose acetate to the process to enhance biological phosphorus removal. Acetate dosing may be required during periods of low organic loading to the anaerobic reactor resulting in reduced volatile fatty acid production required to sustain the phosphorus accumulating bacteria.
- Depending on the final effluent total phosphorus limit, the plant may need to consider chemical coagulant addition to supplement phosphorus removal by chemical precipitation.
- Final effluent solids removal is critical for final removal of the solid phosphorus that is bound up in the cellular structure of the bacterial cells. The plant should follow the steps in this guide for optimizing the solids removal process to ensure maximum phosphorus removal is achieved.
- SOPs for the biological phosphorus removal process should be developed, including an emergency response plan. The emergency response plan should include the following items:
  - Chemical spill containment and cleanup procedures;
  - Procedures for dealing with high nitrate in the sludge recycle to the anaerobic zone. This may be a result of process issues related to the

denitrification process depending on the specific reactor configuration and recycle sludge source; and

- Safety procedures for methanol handling

Phosphorus Removal by Chemical Precipitation

It is common at facilities requiring the removal of phosphorus to use chemical precipitation to achieve the required performance. Biological phosphorus removal can be limited in terms of the amount of removal that can be achieved, and plants that have very low effluent total phosphorus discharge limits often require supplemental removal through chemical precipitation. Some facilities simply would prefer to use chemical precipitation as the only means of phosphorus removal as it eliminates the need for additional biological processes. At smaller facilities, it is often more economically feasible to use chemical precipitation.

In wastewater facilities, the total phosphorus in the secondary treatment mixed liquor will either be bound up in the cellular structure of the aerobic bacteria, or it will be in a soluble form, such as phosphate. Phosphorus that is in the solid form (i.e., bacterial cells) can be removed through physical solids separation prior to effluent discharge. Soluble phosphorus, however, must be converted to a solid form so it, too, can be removed through physical solids separation. This is accomplished by the addition of a coagulant chemical that binds with the soluble phosphate to generate a solid precipitate. The chemical addition is normally performed immediately upstream of the solids separation process. Depending on the effluent discharge limits, a secondary clarifier, or the settling step in an SBR process, may provide sufficient solids removal after chemical precipitation. However, in many cases, a higher level of solids removal is required in a tertiary level treatment process such as filtration or membrane separation.

Chemical coagulants are metal-based chemicals, usually containing iron or aluminum. Some common coagulants used in wastewater treatment include ferric chloride ($FeCl_3$), ferric sulfate ($Fe_2(SO_4)_3$), aluminum sulfate (commonly referred to as alum – $Al_2(SO_4)_3$), and poly-aluminum chloride (PACl). The addition of these chemicals results in the metal ions (Al^{3+} or Fe^{3+}) combining with the phosphate ions in solution to generate a metal-phosphate solid precipitate. Because of the variability of wastewater chemistry, a number of side reactions may occur when chemical coagulant is dosed that do not contribute to the removal of phosphorus. A predominant side reaction can occur between the metal ions and hydroxide ions in solution to generate a metal hydroxide precipitate. These reactions do not result in the removal of phosphorus but they are important to understand as they can contribute significantly to the amount of chemical sludge produced, and can increase the coagulant demand. The chemical equations for these reactions allow for estimation of the theoretical coagulant dosage and the mass of sludge produced, but it is normally recommended to use jar testing with actual wastewater and coagulant samples to optimize the dose for each facility.

Case Study: Achieving Low TP Concentrations using Secondary Activated Sludge Technology

The Composite Correction Program (CCP) and sludge mass control concept was implemented in May 1996 at a 30 MGD secondary activated sludge facility. The CCP addressed a broad spectrum of deficiencies that were limiting the facility from achieving the desired level of performance. Many limitations were nontechnical or management and human-resource related. Such limitations included: inadequate communication between operators and managers, a lack of understanding of facility needs, inadequate application of operational concepts, and inadequate plant coverage to respond to high flow events. The incentive to improve nontechnical skills was based on the realization that any improvements in the secondary effluent TP concentration may not be sustained if leadership and management skills were not developed at the same time. The approach to resolving nontechnical limitations was to address these in conjunction with addressing technical limitations. This approach was highly successful, because it used achievement of optimized performance and capacity as the foundation for change in how utility staff viewed themselves and how they contributed to the organization and their own personal growth. Implementing enhanced communication and applying priority setting and problem-solving skills were absolutely necessary to achieve and sustain the technical changes that were implemented. Gradually, through the numerous process optimization efforts, empowerment of utility staff was dramatically improved.

The secondary unit processes at the plant consist of aeration tanks followed by twelve secondary clarifiers. The facility discharges treated wastewater to a harbor that was experiencing eutrophication, and was a Designated Area of Concern by the International Joint Committee between the U.S. and Canada. The significant reduction in the TP concentration achieved by implementing the total sludge mass control concept is illustrated in Figure 4-16. A target TP of 0.3 mg/L P was adopted to alleviate eutrophication in the harbor. The final effluent target for TP is depicted by the horizontal line on the graph. The monthly average secondary effluent TP concentration is also graphed for comparison with this target. The data displayed is for the period of January 1993 to September 2004. The extended period is provided to illustrate the progressive improvement in performance that occurred as the optimization program matured over several years.

Prior to May 1996 the traditional approach to process control was to only maintain a target concentration of Mixed Liquor Suspended Solids (MLSS) in the aeration tank. There was no measurement or knowledge of the mass of activated sludge in the twelve secondary clarifiers. To help identify and remove all process control bottlenecks, the total sludge mass control concept was implemented in May 1996. Between May 1996 and February 1997, multiple administrative and operations-related performance limiting factors were identified and removed. By February 1997, the activated sludge barrier was fully optimized.

Following the implementation of the total sludge mass control concept, operating staff achieved a level of TP removal with the existing secondary unit processes that was unprecedented. Before optimization, the monthly average final effluent TP concentration ranged from 0.6 mg/L to 1.0 mg/L P. Operating staff first achieved the 0.3 mg/L target in February 1997, ten months after the Composite Correction Program was implemented. The process control stability derived by implementing the total sludge mass control concept resulted in a long-term reduction in the final effluent TP concentration. Eventually the final effluent TP reached a concentration of 0.1 mg/L P in 2003, a level of treatment using secondary technology that was not considered possible by the engineering community. There was considerably less variability in the data from 1997 through 2004, compared to the significant variability that was evident from January 1993 to February 1997. The long-term improvement in the secondary effluent TP concentration was achieved without any design upgrades or major capital expenditure. Utility staff applied a new level of skills in exchange for a new level of plant performance and control. The experience has shown that an intensive focus on identifying and removing process control bottlenecks can turn mediocre performance into impressive performance. Similar assessments of process control and administrative factors are included in chapters 6 and 7, respectively, of this guide.

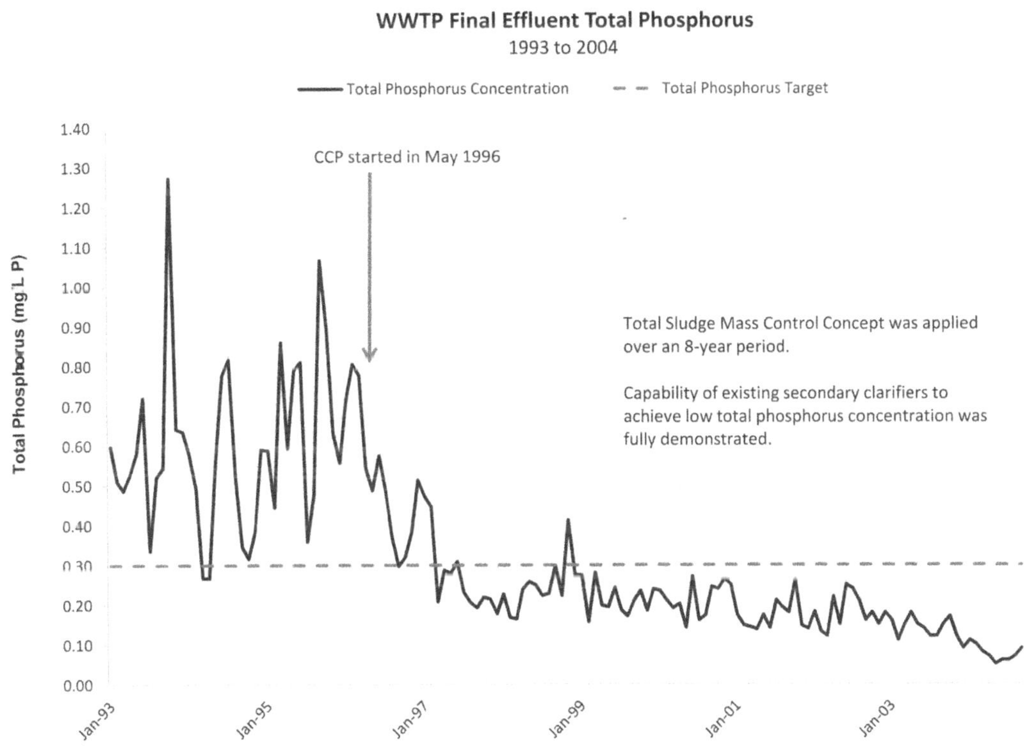

(Source: Gerard Wheeler, Global Facilitation, Inc.)

Figure 4-16. Secondary effluent TP concentration

An additional consideration that must be accounted for is the impact that chemical phosphorus removal can have on the alkalinity concentration. The addition of most of the commonly used coagulants results in the consumption of alkalinity as part of the chemical reactions that take place (both ferric chloride and alum are prime examples of this). If unaccounted for, this can result in a net deficiency in alkalinity at the plant that could reduce the pH below the recommended range. If supplemental alkalinity is being added to the plant, the addition of coagulant for phosphorus removal will increase the demand for alkalinity chemical addition. These simultaneous and somewhat counterintuitive reactions mean chemical addition for phosphorus removal requires a high level of process control, especially at WWTPs that are required to nitrify. The control strategy needs to ensure sufficient chemical is being added to achieve the phosphorus removal targets while avoiding overdosing which leads to excess consumption of alkalinity that may prevent complete nitrification.

Status

The main objective of chemical phosphorus removal is to meet the final effluent total phosphorus limit, with a margin of safety recommended. However, the process of chemically converting soluble phosphate to a metal-phosphate solid precipitate means that a high level of solids control and removal is required to meet the discharge limit for total phosphorus. These two parameters are critical from a monitoring perspective to ensure the process is meeting the stated objectives. In terms of trouble-shooting potential issues with the chemical phosphorus removal system, both the chemical dosing and final solids removal processes would need to be analyzed. Lack of performance could be due to improper chemical dose or solids carryover into the effluent. Operators should review and ensure the solids barriers (e.g., secondary clarifier, tertiary filter, etc.) are optimized in accordance with the applicable sections of this guide.

Once the primary objectives are met and the phosphorus removal is optimized, the operator needs to also ensure that the chemical dosing does not inadvertently impact the biological treatment process by accounting for potential alkalinity impacts. If alkalinity becomes a major concern at the facility, some investigation into the type of chemical coagulant used may prove beneficial. The type of coagulant used can impact the amount of alkalinity consumed. It may be possible that an alternative chemical, such as sodium aluminate, could be used which is a known producer of alkalinity when used in chemical phosphorus removal. A general review of chemical options, complete with jar tests and calculation of the impact of changes on plant operational costs, may prove beneficial in other areas as well in terms of cost savings and possible reduction in chemical sludge production.

The following questions are designed to help the operator better understand the optimization status of the chemical phosphorus removal process. Operators are encouraged to develop and consider additional questions that may be specific to their utility:

- Have performance targets been defined for the system?
  - For the chemical phosphorus removal process, performance targets will be defined by the discharge limits prescribed in the operating permit for total phosphorus, with a safety margin added. In addition, however, many plants that are required to remove phosphorus must also meet stringent effluent TSS concentrations in order to achieve the low phosphorus limit.
- Does the plant have a regular sampling, testing, and monitoring program in place that includes regular testing of the phosphorus removal system performance? Regular testing (daily is recommended) should include the following parameters:
  - Influent and effluent total phosphorus;
  - Influent and effluent total suspended solids; and
  - Effluent alkalinity and pH.
- Do the plant operators conduct regular jar testing using actual wastewater and coagulant samples to verify the chemical dosing system set-points?
- Has the final suspended solids removal system been optimized in accordance with the applicable sections of this guide?
- Does the plant's control system allow for flow pacing of the coagulant chemical to match the wastewater flow? Flow pacing will help to optimize chemical consumption, minimize chemical sludge production, and reduce alkalinity consumption.
- Have the plant operations staff conducted a review of potential alternate chemicals to determine the optimal coagulant?
  - Using several coagulants in parallel jar tests will allow for comparison of key operating parameters such as dosage rate required, chemical sludge production, and alkalinity impacts. In many cases, chemical suppliers will offer assistance with the jar testing or may even provide this as a free service to help with selection of the best chemical for each facility. The plant should be careful when contemplating a switch to a new chemical to ensure the existing dosing and storage equipment materials of construction will be compatible. As an example, ferric chloride is known to be highly corrosive making metal storage tanks unsuitable for chemical storage, and can produce more sludge than other coagulant chemicals.
- Does the chemical storage and dosing system include proper containment for spills to protect the plant staff and the environment? Spill containment pallets or sumps can be used to collect spills or leaks.
- Does the plant have sufficient chemical storage volume to allow for sustained operation in the event of delayed chemical delivery? It would be

recommended to have a minimum of 2 week's storage capacity at average day flow conditions for all process chemicals.

- Have SOPs been developed for the chemical phosphorus removal system that include an emergency response plan? For any chemical systems, the emergency response plan should include chemical spill containment and cleanup procedures in addition to safe chemical handling practices and a copy of the relevant MSDS.

Action

If areas of the chemical phosphorus removal process operation are considered Partially Optimized or Not Optimized, the following steps are recommended for consideration in developing and implementing an action plan.

- Performance targets should be defined for the system, including the effluent total phosphorus and TSS discharge limits.
- A sampling and testing program should be implemented on a daily basis that allows the operator to track phosphorus removal performance and ensure plant alkalinity needs are being met.
- Jar testing should be completed on a regular basis with samples of the chemical coagulant and raw wastewater. Jar testing is normally completed with 500–1,000 mL samples of wastewater and an equivalent dose of the respective coagulant chemical. The goal of the jar test is to simulate conditions in the main plant by matching mixing times, chemical dosage, and effluent solids separation (this can be accomplished by filtering or settling the sample). Jar testing should be completed on a regular basis (e.g., monthly or quarterly would be recommended) or anytime the operator detects a decrease in performance or a change in operating conditions.
- The final effluent solids separation process plays a critical role in removing the metal-phosphate precipitate that has been generated. The solids separation equipment should be optimized in accordance with the respective sections of this guide to ensure optimal phosphorus removal performance.
- If not already in place, the plant should consider implementing a flow-paced chemical dosing system to adjust the chemical feed rate based on the flow of wastewater to the plant. The operator should also adjust the flow pacing settings if the concentration of the total phosphorus in the raw wastewater changes significantly. Not only does flow pacing save money by reducing chemical usage but it also prevents the unnecessary discharge of excess chemical to the environment.

- The plant should consider a review of the available chemicals on the market. Jar testing can be used to evaluate various chemicals under controlled conditions. The plant may discover benefits of using an alternate chemical, including reduced chemical usage, reduced sludge production, positive impact on alkalinity consumption (potential for net alkalinity generation with some chemicals), less hazardous chemical for storage and handling.
- Spill containment should be provided for the plant to protect operator safety and avoid chemical spills to the environment. Some chemicals may also be corrosive in nature and can damage other equipment if spills are not properly contained.
- The plant should have the capability to store several days' worth of chemical as a minimum to ensure chemical is always available in the event of delivery problems or delays.
- SOPs should be developed for the plant that include an emergency response plant.

Performance Limiting Factors

Table 4-5 summarizes factors related to nutrient removal processes that may be limiting performance of the plant. Check whether these factors are Optimized and Documented, Partially Optimized, or Not Optimized. Factors identified as Partially Optimized or Not Optimized will be prioritized in chapter 8, Identification and Prioritization of Performance Limiting Factors/Action Plan Development, where an action plan can be developed and implemented that allows for optimization of these parameters.

Disinfection

This section of the chapter pertains to the disinfection process. The disinfection process is important in that it inactivates potentially harmful microorganisms prior to discharge of the treated effluent into the natural environment. A variety of disinfection options exist, which may be employed in wastewater treatment applications. Two common disinfection techniques, chemical disinfection with chlorine and UV disinfection, are covered in this guide. Plant staff should address the self-assessment questions associated with the relevant process applied at their facility. Facilities using other disinfection techniques are encouraged to similarly evaluate the processes in use at their plant.

Table 4-5. Performance limiting factor summary: nutrient removal

Self-assessment category	Questions for gauging optimization status	Optimization status		
		Optimized and Documented	Partially Optimized	Not Optimized
Biological nitrification	Have performance targets been developed for the nitrification process?			
	Has a sampling, testing, and monitoring program been put into place that includes the required data for nitrification?			
	Have the appropriate trend charts been developed to collect the required data and track actual system performance relative to the defined operating targets?			
	Has a maintenance and inspection log been developed for the plant that includes the secondary treatment system and nitrification process?			
	Have SOPs been developed for the nitrification process, including an emergency response plan?			
Biological denitrification	Have performance targets been developed for the biological denitrification system?			
	Has a sampling and testing program been developed that includes the denitrification process?			
	Have all data been plotted on trend charts along with applicable performance targets?			
	Has the nitrification process been optimized in accordance with the sludge mass control concept?			
	Has a nitrate mass balance been completed for the pre-anoxic denitrification process to ensure sufficient recycle of nitrate from the aeration tank?			
	If the plant includes a post-anoxic denitrification step, has the pre-anoxic tank been optimized for maximum nitrate removal? Has the appropriate testing and data collection been implemented to verify nitrate removal in each anoxic zone?			
	If post-anoxic denitrification is used, has the supplemental carbon dosing system been optimized for chemical usage efficiency?			
	If methanol is used for supplemental carbon dosing, have alternative, less hazardous chemical options been explored?			

Self-assessment category	Questions for gauging optimization status	Optimization status		
		Optimized and Documented	Partially Optimized	Not Optimized
Biological denitrification *(continued)*	Has a routine maintenance and inspection program been developed that includes the denitrification process?			
	Have SOPs been developed for the denitrification process, including an emergency response plan?			
Biological phosphorus removal	Have performance targets been defined for the biological phosphorus removal process?			
	Has a total mass control program been implemented to define a target sludge mass and SRT that balances the requirements for biological phosphorus removal with total nitrogen removal?			
	Has a sampling, testing, and monitoring program been developed for the biological phosphorus removal process?			
	Has the plant optimized the acetate chemical dosing system to maximize phosphorus removal and maintain a consistent mass of phosphorus accumulating bacteria?			
	Does the plant have the capability to supplement a chemical coagulant to enhance phosphorus removal? If so, has this system been optimized according to the chemical phosphorus removal section of this guide?			
	Has the final effluent solids removal process been optimized in accordance with the applicable section(s) of this guide?			
	Have SOPs been developed for the biological phosphorus removal process that include an emergency response plan?			
Chemical phosphorus removal	Have performance targets been defined for the system?			
	Does the plant have a regular sampling, testing, and monitoring program in place that includes regular testing of the phosphorus removal system performance?			
	Do the plant operators conduct regular jar testing using actual wastewater and coagulant samples to verify the chemical dosing system set-points?			

Continued

Self-assessment category	Questions for gauging optimization status	Optimization status		
		Optimized and Documented	Partially Optimized	Not Optimized
Chemical phosphorus removal *(continued)*	Has the final suspended solids removal system been optimized in accordance with the applicable sections of this guide?			
	Does the plant's control system allow for flow-pacing of the coagulant chemical to match the wastewater flow?			
	Have the plant operations staff conducted a review of potential alternate chemicals to determine the optimal coagulant?			
	Does the chemical storage and dosing system include proper containment for spills to protect the plant staff and the environment?			
	Does the plant have sufficient chemical storage volume to allow for sustained operation in the event of delayed chemical delivery?			
	Have SOPs been developed for the chemical phosphorus removal system that include an emergency response plan?			

Chemical Disinfection with Chlorine-Based Chemicals Understanding

Chlorine addition has historically been the most common method of disinfection for both water and wastewater treatment. This is largely due to the widespread availability of chlorine chemicals and its proven effectiveness as a disinfectant. Several different chlorine-based chemicals have been used for disinfection purposes, including chlorine gas, liquid chlorine, sodium hypochlorite, calcium hypochlorite, and chlorine dioxide. Historically, liquid or gaseous chlorine was used most commonly but these chemicals are toxic and storage and handling requirements have become increasingly stringent due to potential impacts on the environment. Many plants have opted to switch to alternative chemicals such as sodium or calcium hypochlorite that are less hazardous from a storage and handling perspective, or have chosen to switch to on-site generation of sodium hypochlorite.

Chlorine is effective in destroying disease-causing microorganisms. One of the challenges with dosing chlorine is that a number of side reactions occur with various oxidizable materials commonly present in treated wastewater. For example, organic material, ammonia, and various metal ions all exhibit a chlorine demand and can readily react with chlorine to form compounds such as chloramines or other disinfection by-products. The generation of disinfection by-products adds to the total chlorine demand required for disinfection. The disinfection by-products and other organo-chlorine compounds formed during the disinfection process must be oxidized before the destruction of disease-causing organisms is achieved. Because of this, breakpoint chlorination methodology is typically used to ensure sufficient chlorine has been dosed to the system. When chlorine is added to treated wastewater, it will react to satisfy all chlorine demand prior to its availability as free chlorine. Free chlorine will only exist after sufficient chlorine has been added to react with all oxidizable substances in the effluent. The presence of free chlorine in the effluent flow is referred to as chlorine residual. Maintaining a free chlorine residual downstream of the disinfection process ensures that enough chlorine has been added to satisfy all the side reactions as well as meet the disinfection requirements for disease-causing organisms.

Chlorine dosing for disinfection must also consider the contact time provided between the injection point and the final discharge location. Sufficient time and thorough mixing must be provided to ensure proper disinfection. This is typically achieved through the application of the CT concept. The CT concept is the product of the dosage concentration (C) multiplied by the contact time (T) available. Contact time is a function of the contact chamber design and the effluent flow rate. If the contact chamber is undersized and is not capable of providing sufficient contact time at the given concentration, the plant may be able to satisfy the CT requirement by increasing the chemical concentration used for disinfection. Proper mixing of chlorine chemical is also critical to achieve the desired disinfection performance.

An additional concern with chlorine is its toxicity in the receiving environment. Discharge of free chlorine to the environment can result in the formation of toxic compounds in the receiving water body, which can be harmful to organisms residing in that habitat. Where chlorine is used for disinfection, the free residual must be removed through a process known as dechlorination. Reducing agent chemicals, such as sulfur dioxide or sodium bisulfite, are typically dosed at the end of the chlorine contact chamber to react with the free chlorine residual and form nontoxic compounds for discharge to the environment. Often discharge permit limits reflect the requirements for dechlorination. Permit limits for total residual chlorine for plants that dechlorinate can often be very low. It is important that instrumentation and methods used for monitoring effluent concentrations of chlorine residual has a detection limit sufficiently low to determine compliance. This concept is discussed in greater detail in chapter 6.

Status

The main objective of the chlorination process is the destruction of disease-causing microorganisms for public health protection, and is a primary focus of the operator. The operator must carefully manage the operating CT value (dosage concentration multiplied by the contact time) for the system to ensure the proper level of disinfection is achieved. Since the operator may have little control over the time component of the CT value due to contact chamber design limitations and plant flow rates, the dosage of chlorine becomes an important process control parameter. The challenge for optimizing a chlorine disinfection system involves ensuring sufficient chemical dosage at all times without overdosing. Overdosing of chlorine chemical is wasteful of plant resources, and increases the demand for chemicals required for dechlorination.

The following questions are designed to help the operator better understand the optimization status of the chlorine disinfection process. Operators are encouraged to develop and consider additional questions that may be specific to their utility:

- Have performance targets been defined for the chlorine disinfection system?
    - Operating targets will typically be defined in the operating permit for the plant and will include the allowable total and/or fecal coliform count (usually measured in MPN/100 mL) and the allowable chlorine residual downstream of the dechlorination process.
- Has a defined CT value been set for the plant to achieve the required level of disinfection? Original design documentation for the plant may assist in selecting a target CT. In the absence of a design value, numerous publications provide CT tables that include recommended values to achieve the target disinfection removals.
- Has a regular sampling and testing program been implemented at the plant that includes analyzing the final effluent for disinfection system performance?
    - Some level of testing will be required by the operating permit or government regulations. Additional sampling and testing will help to supplement the required testing to better define operating trends for the plant. It is important that the testing methodology and instrumentation be of sufficient sensitivity to determine compliance, particularly for plants with extremely low discharge permit requirements for total residual chlorine.
- Does the plant's control system allow for flow-paced chlorine dosing to allow for reduced chemical usage at lower flow rates?
- Does the plant have instrumentation that allows for automated control of the dechlorination chemical dosing pumps?
- Has the plant reviewed the various chlorine-based chemical options available for disinfection? The use of sodium hypochlorite for example as opposed to liquid or gaseous chlorine may offer benefits in terms of handling and operator safety.

- Has the plant reviewed nonchlorine based chemicals as alternatives to chlorine? Chlorine is recognized for its toxicity and has faced increased scrutiny as a disinfectant. Alternative chemicals such as ozone, UV, and peracetic acid may offer the potential benefits of reduced toxicity concerns, differing for handling and storage requirements, and elimination of the dechlorination step.
- Has a regular maintenance and inspection program been developed for the plant that includes the chlorine disinfection system?
- Does the plant have sufficient chemical storage to deal with minor delays or disruptions to delivery of new chemical?
- Have SOPs been developed for the chlorine disinfection system, including an emergency response plan? The emergency response plan should include, as a minimum, how to respond to high levels of coliform counts exceeding the permit limit and a chemical spill response plan.

Process Energy Considerations

The main opportunity to save energy in the chlorine disinfection system is by reducing the chlorine and dechlorination chemical dosages during periods of low plant flow. This can be accomplished by using online instrumentation to match the actual dosage rate to the demand. In addition to achieving reduced chemical costs, there may also be a small savings in electricity by running dosing equipment at reduced speeds.

To assess the energy efficiency of the chemical disinfection process, the following self-assessment question should be considered:

- For chemical disinfection (chlorine and nonchlorine), is the chemical dosage rate automatically adjusted to meet the demand?

Action

If areas of the chlorine disinfection system operation are considered Partially Optimized or Not Optimized, the following steps are recommended for implementing an action plan:

- Performance targets should be clearly defined for the system. The targets would typically include the effluent total and/or fecal coliform limits and allowable free chlorine residual.
- A target CT should be defined for the plant based on the required coliform log removal. The target CT should not be used as a guarantee of disinfection and should be used as a guide to implement the initial control points (i.e., contact time and free residual) of the disinfection system. Dosing requirements should always be based on actual coliform removal

performance. The operator can also use the CT concept as a guide for troubleshooting when design limitations are present within the chlorine contact chamber. Increasing the dosage concentration may overcome less than optimal chlorine contact time. Low contact time can also be a result of short circuiting within an improperly designed contact chamber. It may be possible to improve the flow pattern through the contact chamber by installing additional baffles to create a more serpentine configuration.

- That plant should implement a regular sampling and testing program that meets the requirements of the operating permit. It is recommended to implement testing as often as possible which will help to better define trends in the data and may help with optimizing chlorine dosage rates and control system set-points. Ensure that instrumentation and analytical methods used have a method detection limit sufficiently low to determine compliance.
- If not already in place, the plant should consider a flow-paced dosing system that allows for reduced chemical addition during periods of low plant flow. The operator will need to carefully monitor the system to ensure the set-points achieve the required level of disinfection at all times.
- At some plants, addition of dechlorination chemical, such as sulfur dioxide, may also be accomplished using flow-pacing similar to the chlorine addition. Further optimization may be possible by using online free chlorine residual measurement to control the dechlorination dosage.
- The plant should consider reviewing alternate chemical options that may be less hazardous and/or toxic compared to liquid or gaseous chlorine. Alternate chemicals, such as sodium hypochlorite, are still chlorine based but tend to be considerably less hazardous. Other nonchlorine-based chemicals, such as peracetic acid, ozone, and UV, offer a further benefit of eliminating the need for a dechlorination step. A review of chemicals should include a life-cycle cost estimate, review of handling and safety issues, compatibility of existing dosing system materials of construction, and availability of the alternate chemical.
- A regular maintenance and inspection program should be developed for the chlorine disinfection system. Operators should regularly inspect the chlorine dosing system for any signs of leaking that create hazardous or toxic conditions. Operators should also regularly inspect the chlorine contact chamber for biofilm growth or other debris that may impact performance.
- The plant should review chemical storage available to ensure it is adequate. It is recommended to have a minimum of 2 weeks' chemical storage to allow for continues operation in the event of delays or short-term disruptions in chemical supply.

- SOPs should be developed, including an emergency response plan and chemical spill containment and cleanup procedure.

Disinfection with Ultraviolet (UV) Radiation

Understanding

The use of ultraviolet (UV) light has gained widespread use in the industry for the disinfection of wastewater. Typically, low-pressure mercury-filled bulbs are used to generate UV light at wavelengths in the range 185–254 nm. These wavelengths have been proven effective for deactivating viruses and bacteria. It is most common in the industry to find UV lights submersed in a free-flowing effluent channel. The UV bulbs are typically housed in quartz sleeves and emit the UV radiation into the wastewater as it flows through the channel. An alternative configuration, allows the wastewater to flow within clear tubes while the UV bulbs are external to the flow tubes and remain dry. In some cases, the UV bulbs may be supplied within a pressure vessel for installation in pressurized pipe systems as opposed to using an open channel. UV may be used in conjunction with a chemical compound, such as hydrogen peroxide, where additional disinfection strength or the oxidation of recalcitrant chemical compounds is required.

Regardless of the configuration, the effectiveness of the UV disinfection process is highly dependent upon the UV light intensity output from the bulbs (measured in Watts), the effluent flow rate and the quality of the effluent. An effluent that is low in total suspended solids, turbidity, and organics (UV-absorbing organics) will usually result in more efficient disinfection and will typically require less light output for a given flow rate. The presence of solids particles, high turbidity, or UV absorbing organic substances may interfere with the proper transmission of the light waves passing through the water. The most effective and accurate means of measuring the quality of the water being disinfected is to measure the UV transmittance at 254 nm. Online analyzers and benchtop or portable spectrophotometers are available to measure UV transmittance in real time. These instruments work by passing a known UV intensity through a small sample of the effluent to measure how much of the light passes through the sample, relative to a deionized water blank. The result from the test is the UV transmittance value, measured directly as a percentage. A higher percentage measurement indicates a sample that is of high quality, as most of the UV light is able to pass freely through the sample. Often these types of measurements are used when designing and selecting the UV equipment for the facility but regular measurements can help identify potential problems with the upstream process that may be leading to abnormal UV transmittance values and poor disinfection system performance. The UV disinfection process relies on delivering a certain light intensity to all molecules in the flowing effluent stream. Any factors that may impact on this capability will have a negative impact on the performance, including scale or biofilm growth on the quartz sleeves. For this reason, some UV systems are outfitted with wiper systems which may include chemical sprays for removing any buildup (such as precipitated chemical coagulant) on the sleeves. Properly maintaining the integrity of the UV bulbs and sleeves is an important part of ensuring process performance.

Status

For the operations team, the performance of the UV disinfection system is commonly monitored in terms of the pathogen concentration of the final effluent after disinfection. Samples are typically analyzed for fecal and total coliform concentrations. These concentrations are typically measured in units of MPN/100 mL and usually one or both of these parameters appears in the facility's effluent discharge permit. This analysis is most commonly performed by an external laboratory as operating permits generally require analysis for regulatory reporting to be completed by an accredited laboratory. It is recommended that coliform concentrations also be tested upstream of the disinfection process to allow calculation of the percentage inactivation produced by the process. While the analysis of secondary effluent (predisinfection) and final effluent (postdisinfection) coliform concentrations measures the performance of the system, it is not considered a good analysis tool for process monitoring due to the time lag in sample collection and confirmation of the results. Often the test results are not available for several days after the collection of the sample. The operator is encouraged to use additional testing methods to monitor the operation and performance of the UV disinfection system.

For example, most UV systems include the capability to monitor the lamp output, often identified as UV intensity, measured in Watts. Additionally, portable test kits or online analyzers are available for quick analysis of the UV transmittance of the final effluent being disinfected. These properties both can impact significantly on the effectiveness of the UV disinfection process and careful monitoring of these parameters could provide an early indication of performance issues that may lead to decreased pathogen inactivation. If the UV output intensity drops below design values, this is an indication of problems with the system itself and may be indicative of a lamp failure or reduced performance due to lamp age. Reduced UV intensity is a signal to the operator to inspect the lamps and consider replacement of faulty or aged lamps. Reduced UV transmittance is a property of the wastewater system effluent and may indicate reduced performance from the upstream biological treatment process or abnormal contaminants in the source wastewater. Properties such as increased effluent TSS, increased turbidity, and increased color can potentially impact on the UV transmittance value. Plotting these values on a trend chart will allow the operator to visualize changes over time. For example, as the lamps age, it is natural for the UV intensity to decrease slowly over time. If the operator detects a sudden decrease in intensity output that does not seem to be in-line with the long-term trend, it could be indicative of a problem that needs to be addressed such as a premature lamp failure or extensive fouling.

The following questions are designed to help the operator better understand the optimization status of the UV disinfection process. Operators are encouraged to develop and consider additional questions that may be specific to their utility:

- Have performance targets been developed for the UV disinfection system? These targets impact directly on the quality of the effluent being discharged and are typically defined by the regulatory agency and included in the operating permit for the facility.
    - ▲ Typically, performance targets for a disinfection system are based on effluent coliform bacteria counts. The permit may be based on fecal coliform, total coliform, or both and are typically measured in units of MPN/100 mL.
- Does the operator have the capability to monitor key operating parameters for the UV disinfection system, including UV intensity and effluent UV transmittance?
    - ▲ UV intensity is a property of the disinfection system itself and is an indication of the output strength of the UV radiation generated by the lamps. It is common for the UV intensity to decrease slowly over time as the lamps age and this is usually built into the design of the UV system. The operator should be concerned if UV intensity decreases below design values for the system.
    - ▲ UV transmittance is a property of the effluent being disinfected and indicates the percentage of UV light that is able to pass through the sample. Portable test kits can be purchased that allow fast and easy measurement of the transmittance value for a sample. Typically, UV systems are designed for a specified minimum transmittance (often identified as UVT).
- Have trend charts been developed for the UV system to monitor key operating and performance parameters? Recommended parameters to be monitored include influent and effluent coliform concentrations, UV intensity, and UV transmittance.
- Has an inspection and logging program been developed for the plant that includes the UV disinfection system?
- Have SOPs been developed for the UV disinfection system, including maintenance procedures and an emergency response plan?

Process Energy Considerations

The main source of energy consumption in a UV disinfection system is electricity required to power the ultraviolet bulbs. Some UV systems, depending on the manufacturer and the model, allow for varying the number of bulbs receiving power at any time. This control is usually based on the flow rate or the level measured in the UV channel. Some systems may allow for bulbs to be switched off when the liquid level drops in the channel. Other configurations may switch off entire banks of bulbs when the flow rate is below a certain set-point. Any system that uses this type of control must always be carefully monitored to ensure the set-points are still providing the required level of radiation for disinfection as bulbs are switched off. Depending on the size of the system, this type of control could result in a significant electricity savings on an annual basis. Smaller plants will likely notice little impact on total plant electricity usage, and the ability to switch bulbs off may not even be available as an option on smaller systems.

To assess the energy efficiency of the UV disinfection process, the following self-assessment question should be considered:

- For UV disinfection, does the system have the capability to vary the number of lamps in service based on UV transmittance or flow or level in the UV channel?

Action

If areas of the UV disinfection system operation are considered Partially Optimized or Not Optimized, the following steps are recommended for implementing an action plan.

- Performance and operating targets for the disinfection system should be developed, including the effluent coliform concentration limit, minimum recommended operating UV intensity, and the minimum recommended UV transmittance. The operators may need to consult original design documents to determine what the system was originally designed for.
- It is recommended to develop a regular sampling, testing, and monitoring program to regularly track and analyze the process performance. The monitoring program should include the following parameters:
  - Disinfection system influent and effluent coliform concentrations: Note that only the effluent coliform concentration is normally required for regulatory reporting but testing of the influent concentration allows for calculation of pathogen deactivation in the UV system.
  - UV intensity: Note that many UV systems are designed to continuously monitor UV output and often will alarm on low intensity.
  - UV transmittance should be monitored using an online analyzer or laboratory/portable UV spectrophotometer: This parameter is an indicator of the water quality being disinfected. Often UV systems are designed for a minimum UV transmittance value. It is also recommended to measure turbidity as an estimate of the amount of suspended solids present in the sample.

- Trend charts should be developed to track the data and to detect abnormal operation in the UV disinfection process. Some recommendations for follow-up action based on abnormal trends would be as follows:
  - ▲ If low UV intensity is detected, it is indicative of a problem with the UV system itself. This could be attributed to aged lamps or a sudden failure of one or more lamps/modules in the system.
  - ▲ If low UV transmittance is detected, this is a signal that the water quality flowing into the UV system has decreased. This could be a result of issues with the upstream treatment process (an example would be excessive solids carryover from the secondary clarifier) or possibly changes in the raw wastewater quality.
- An inspection and logging program should be developed for the entire plant to include the UV disinfection process. Some key items that should be routinely checked as part of a visual inspection of the UV system include the following:
  - ▲ Regular inspection of the UV lamps for scale formation or biofilm growth that may be reducing the effectiveness of the UV radiation.
  - ▲ Visual inspection of the flow channel for any debris or other buildup.
  - ▲ Observation of the flow path through the UV system for smooth flow. Any potential issues within the flow channel that may be creating abnormal flow currents, short-circuiting, dead zones, and/or uneven distribution in the channel could lead to ineffective disinfection.
- SOPs should be developed for the UV disinfection system, including maintenance procedures and an emergency response plan. Some items that should be considered in the emergency response plan are as follows:
  - ▲ Provision for backup power;
  - ▲ Spare lamps/modules on hand for sudden or premature failures; and,
  - ▲ Operational considerations such as switching to alternate or standby units.

Performance Limiting Factors

Table 4-6 summarizes factors related to the disinfection process that may be limiting performance of the plant. Check whether these factors are Optimized and Documented, Partially Optimized, or Not Optimized. Factors identified as Partially Optimized or Not Optimized will be prioritized in chapter 8, Identification and Prioritization of Performance Limiting Factors, where an action plan can be developed and implemented that allows for optimization of these parameters.

Table 4-6. Performance limiting factor summary: disinfection

Self-assessment category	Questions for gauging optimization status	Optimization status		
		Optimized and Documented	Partially Optimized	Not Optimized
Chlorine disinfection	Have performance targets been defined for the chlorine disinfection system?			
	Has a defined CT value been set for the plant to achieve the required level of disinfection?			
	Has a regular sampling and testing program been implemented at the plant that includes analyzing the final effluent for disinfection system performance?			
	Does the plant's control system allow for flow-paced chlorine dosing to allow for reduced chemical usage at lower flow rates?			
	Does the plant have instrumentation that allows for automated control of the dechlorination chemical dosing pumps?			
	Has the plant reviewed the various chlorine-based chemical options available for disinfection?			
	Has the plant reviewed nonchlorine based chemicals as alternatives to chlorine?			
	Has a regular maintenance and inspection program been developed for the plant that includes the chlorine disinfection system?			
	Does the plant have sufficient chemical storage to deal with minor delays or disruptions to delivery of new chemical?			
	Have SOPs been developed for the chlorine disinfection system including an emergency response plan?			
UV disinfection	Have performance targets been developed for the UV disinfection system?			
	Does the operator have the capability to monitor key operating parameters for the UV disinfection system including UV intensity and effluent UV transmittance?			
	Have trend charts been developed for the UV system to monitor key operating and performance parameters?			

Self-assessment category	Questions for gauging optimization status	Optimization status		
		Optimized and Documented	Partially Optimized	Not Optimized
UV disinfection *(continued)*	Has an inspection and logging program been developed for the plant that includes the UV disinfection system?			
	Have SOPs been developed for the UV disinfection system, including maintenance procedures and an emergency response plan?			
Process energy considerations	For chemical disinfection (chlorine and nonchlorine), is the chemical dosage rate automatically adjusted to meet the demand?			
	For UV disinfection, does the system have the capability to vary the number of lamps in service based on UV transmittance or flow or level in the UV channel?			

Advanced/Tertiary Treatment

This section of the chapter pertains to advanced/tertiary treatment processes. Although a variety of advanced treatment processes are employed, for both wastewater treatment and reuse applications, a limited number of these processes are addressed in this section, including tertiary filtration and membrane bioreactors. Plant staff are encouraged to consider the self-assessment questions for relevant processes employed at their facilities. For processes not included in this guide, general self-assessment guidance for additional processes is included at the end of this chapter.

Tertiary Filtration

Understanding

Tertiary filtration is used in plants that require a higher level of suspended solids removal that cannot be achieved by secondary clarification alone. Tertiary filtration may be installed to achieve a low total phosphorus limit as well. Biological and chemical phosphorus removal involves converting soluble phosphate to a solid form. Tertiary filtration may be required if the size of the phosphorus precipitant is small and cannot be settled in a secondary clarifier.

Tertiary filtration refers to the physical removal of suspended solids by passing the effluent through a bed of granular material. The most common filter bed media is sand but other materials, such as anthracite, granular activated carbon, or synthetic media, may be used. Granular activated carbon is a media that may be used to absorb specific organic compounds from the wastewater stream prior to discharge. Examples of these compounds may include pharmaceutical compounds, their metabolites, or other contaminants of emerging concern, which may have the potential to be detrimental to the natural environment or a downstream drinking water source.

Passing the effluent through the filter traps the particulate material in the filter bed while the clarified liquid passes through. Filters may be configured as either up-flow or down-flow, and a wide range of hydraulic configurations can be used. Once a filter's capacity is exhausted, filters are typically backwashed by passing water back through the filter to remove particulate matter. Backwash is a critical filter operation technique, and it is important that parameters be established regarding the initiation of backwash, determining backwash duration, and returning filters to service after backwash has been completed. Regardless of the specific configuration, the general principle of solids removal remains the same for media filtration techniques. A complete discussion of filter optimization is beyond the scope of this guide. Additional filtration optimization information may be accessed in a variety of industry resources.

Status

Tertiary filtration primarily relies on physical filtration for removal of suspended solids. In filters where biomass is allowed to grow, microorganisms may also contribute to the efficacy of the filtration process by breaking down specific contaminants in the influent water. Regular inspection and maintenance of the filter unit is the best method for maintaining optimal solids removal. Over time, as the filters are operated, they will become plugged with the solids trapped on the media. Typically backwashing with water or cleaning with compressed air is required to remove the solids material from the filter bed and restore its performance. Filter cleaning or backwashing may be initiated manually by the operator if high suspended solids are detected in the effluent, or it may be initiated automatically if high headloss is caused by clogging. It is important to define the parameters associated with the initiation and conclusion of backwash, as described previously. It is also helpful to regularly assess the physical condition of media filters through basic filter surveillance techniques.

The following questions are designed to help the operator better understand the optimization status of the tertiary filtration process. Operators are encouraged to develop and consider additional questions that may be specific to their utility:

- Have performance targets been defined for the tertiary filtration process?
- Has a sampling and testing program been implemented for the tertiary filtration process that allows actual performance to be tracked and compared to the defined targets?
- If chemical phosphorus removal is included at the plant, has the chemical coagulant dosage been optimized for use with the tertiary filter?
- Is there a provision for continuous operation during filter cleaning and/or backwashing operations?

- Have backwash initiation and return-to-service parameters been defined?
- Does the control system have the ability to detect and automatically initiate a backwash?
- Has a regular maintenance and inspection program been developed for the tertiary filters?
- Have SOPs been developed for the tertiary filter, including an emergency response plan?

Action

If areas of the tertiary filtration system operation are considered partially optimized or not optimized, the following steps are recommended for consideration in developing and implementing an action plan.

- Performance targets should be defined for the process. Performance targets would include the maximum solids loading to the filter and the effluent suspended solids concentration required by the operating permit, as well as any other parameters of interest. If the filter is regularly being overloaded, the only solution may be to upgrade the system or add additional units.
- A sampling and testing program should be implemented for the tertiary filtration system. Samples of the filter influent and effluent should be collected daily and analyzed for TSS concentration. This will allow for calculation of the solids loading on the filter as well as removal rates. If granular activated carbon filtration is used for the removal of a specific contaminant compound, the filter effluent should be periodically analyzed for the parameter of interest.
- If chemical phosphorus removal is included at the plant, the operators should carefully evaluate the optimal coagulant dosing location for the process. Dosing chemical coagulant upstream of the secondary clarifier will result in a reduced solids loading on the tertiary filter but increased chemical sludge removed as WAS from the clarifier. Conversely, dosing between the secondary clarifier and the tertiary filter will reduce the chemical sludge in the WAS but will increase the solids loading on the filter. The filter must be properly designed to handle the additional loading if staff decide to remove chemical phosphorus precipitant.
- It is advantageous for filter units to be equipped with online instrumentation to detect high head loss conditions through the filter, as well as high turbidity in the effluent water, and automatically initiate a backwash event as required based on the plant's defined backwash initiation parameters.

- A regular maintenance and inspection program should be implemented for the tertiary filters. Daily inspection of the filters should be completed to ensure proper operation. The presence of odors or biofilm growth within the filter may indicate a need for filter cleaning. Documentation provided by the filter manufacturer should identify regular maintenance and cleaning procedures. Consider instituting more detailed filter surveillance practices to document long-term filter condition and performance over time.
- SOPs should be developed for the system, including an emergency response plan. The emergency response plan should include provisions in the event a filter must be taken out of service for an extended period of time (i.e., filter media replacement).

Membrane Bioreactor (MBR) Process

This guide shows that one of the most important parameters impacting the performance of a wastewater treatment system is solids control. Within the activated sludge system, controlling the SRT is of paramount importance for process stability, particularly when nitrification is required. For total phosphorus removal, either by biological or chemical processes, the removal of suspended solids from the final effluent is critical. In many locations, effluent TSS limits are becoming more stringent as environmental protection becomes a much higher priority than in the past. In many cases, these requirements cannot be met by conventional treatment systems and alternative, advanced technologies have been developed to fill this need. Membrane treatment systems have been developed based on providing a physical barrier that virtually eliminates suspended solids in the effluent. In some cases, the membranes have been used as a post-secondary tertiary filter but it is most common to find the membranes immersed in the secondary mixed liquor in what has become known as a membrane bioreactor (MBR) system. In either case, the membranes have a very fine pore size. Most commercially available membranes have a pore size less than 1 micron, a size that is smaller than a bacterial cell. Nearly all of the TSS in the wastewater is retained within the system, providing a very high quality of effluent.

MBR systems offer additional advantages in addition to a virtually TSS-free effluent. Because the membranes provide virtually 100 percent retention of solids within the system, there is almost no opportunity for the loss of precious biomass due to solids washout. Where secondary clarifiers can be impacted negatively by high MLSS concentrations in the aeration tank leading to incomplete settling, the membranes still provide a physical barrier and do not rely on gravity settling for solids separation. MBR systems, therefore, can typically operate at much higher MLSS concentrations. The MLSS in an MBR system can approach or even exceed 10,000 mg/L depending on the membrane manufacturer and the specific design conditions for the facility. The net impact is that a smaller aeration tank volume is required to provide the same biomass. Overall, this reduces the system footprint, making it particularly attractive for smaller sites or retrofit plants. The use of membranes also alleviates many of the concerns with conditions that lead to poor settling sludge. In fact, poor settling sludge has virtually no consequence in an MBR system as the membranes will always retain the solids in the aeration zone.

In recent years, the use of membranes has become increasingly economical and more widespread in wastewater treatment. There are now many membrane manufacturers with unique materials, configurations, and operating requirements. In general, however, the commercially available membranes can be classified as either flat-sheet style or hollow fiber strands. Flat-sheet style membranes typically have two layers of membrane material fused to an internal rigid structure that is commonly referred to as a cartridge. During treatment, the effluent (commonly called permeate for membrane systems) flows across the membrane sheets into the inner space within the cartridge. Cartridges are usually connected to a main permeate header that leads to the final discharge. Hollow fiber strand membranes are tubular in shape with many strands connected to a common permeate header. Permeate flows across the membrane surface into the inner volume within each strand before accumulating in the header pipe. In either membrane configuration, flow is often generated with a pump on the downstream permeate header although in some rare cases flow can be generated by gravity if the appropriate hydraulic conditions allow for it.

Within an MBR system, the membranes themselves are submerged within the mixed liquor and permeate is drawn across the surface. Mechanical cleaning is often required for the membranes to reduce the accumulation of a dense cake layer on the mixed liquor side of the membrane surface. It is most common to use diffused aeration to create a scouring flow of air bubbles across the entire membrane surface to help reduce cake formation that can lead to membrane fouling (i.e., significant blocking of the membrane pores). The aeration provides oxygen for the aerobic bacteria while also maintaining good permeability of the membranes. However, in most membrane systems, the scouring action is not sufficient to permanently maintain permeability and the membranes may still become fouled with both organic solids (e.g., biomass) and inorganic solids (e.g., scale). Most membrane systems are therefore supplied with one or more chemical cleaning systems that typically backflush through the permeate lines to fill the inner volume of the membranes. Common chemicals used for membrane cleaning include citric acid and sodium hypochlorite. Maintaining membrane permeability within an MBR system is key for successful operation, and this can be achieved through determining and establishing an optimal cleaning frequency and protocol for the membrane system.

Status

Most membrane systems are designed for a maximum flow of wastewater per unit of membrane surface area in the system. This parameter is defined as the membrane flux rate, and each membrane manufacturer will have different theoretical maximum flux rates that their systems are designed for. Monitoring the rate of permeability loss in the system over time is typically done by monitoring the transmembrane pressure (TMP). As permeate is drawn across the membranes, a certain differential pressure is required either by pumping or by gravity to generate a flow from the outside of the membrane (mixed liquor side) to the inside (treated permeate side). As sludge cake begins to build on the membrane surface and the membrane becomes fouled, it becomes harder to draw permeate through the membranes and a larger TMP is required to generate flow. It is common for membrane manufacturers to design for a maximum TMP allowed for their system. When the TMP approaches this limit, it is a signal to the operator that the membranes require chemical cleaning to restore permeability.

Since an MBR process is, at its core, still a suspended growth activated sludge process, many of the same operating parameters still apply. Parameters, such as F:M ratio and OLR, are still useful for MBR systems. Because of the high concentration of biomass in the system it is common for MBRs to operate at much higher OLRs than the CAS process. In many MBR systems, a conventional aeration tank is used to perform the bulk of the BOD oxidation and nitrification processes. Often the membranes are installed in smaller, separate membrane tanks downstream from the aeration tanks. In this type of configuration, the MBR tank is analogous to a secondary clarifier in that its primary purpose is to remove solids from the effluent and return the active biomass to the aeration tank via RAS pumps and piping. Operators must still apply the total sludge mass control concepts to membrane systems and include the entire biomass (i.e., aeration tank + membrane tank) in the SRT calculation and sludge wasting control strategy. The primary difference versus CAS systems is that MBRs are often designed for a longer SRT, which must be taken into account in the calculations.

The following questions are designed to help the operator better understand the optimization status of the membrane bioreactor (MBR) process. Operators are encouraged to develop and consider additional questions that may be specific to their utility:

- Have performance targets been defined for the process?
    - Performance targets would include effluent treatment standards as defined in the operating permit as well as target SRT and total sludge mass targets. These targets would be similar to those derived for the CAS process. Additional performance targets for the membranes would include maximum flux rate and transmembrane pressure. These parameters would typically be defined by the membrane manufacturer in the design documentation.
- Has a sampling and testing program been developed for the membrane process that allows tracking of actual system performance relative to the defined targets? The membrane process is capable of BOD_5 and TSS removal and may also be part of a biological nutrient removal system for

nitrogen and phosphorus. Operators should consult the applicable sections of this guide for further recommendations on sampling and testing for nutrient removal.

- Has the total sludge mass control program been adapted and applied to the membrane bioreactor process?
  - ▲ It is common for a conventional aeration tank to be followed in series by the membrane tank. For application of the mass control program, the membrane tank is analogous to a secondary clarifier in that its primary function is solids removal and controlling the biomass inventory and distribution using the RAS and WAS systems.
- Does the plant control system actively monitor the membrane permeability? Typically, this is done through online measurement and feedback of the transmembrane pressure.
- Does the system include provision for restoring the permeability of the membranes when they have become fouled? This may be accomplished through mechanical means (i.e., scouring) or through periodic chemical cleaning.
- If chemical cleaning is used, has a cleaning frequency and protocol been established?
- Does plant have an effective method of controlling foam events? Foaming can be an issue with MBRs due to the relatively high organic loading compared with other biological processes.
- Have SOPs been developed for the process that include an emergency response plan?

Process Energy Considerations

The membrane bioreactor process is a variation on the activated sludge process and may also be capable of nutrient removal depending on the design requirements. Operators are encouraged to review the process energy considerations identified for the CAS process as well as applicable nutrient removal section for potential application to the membrane bioreactor.

Action

If areas of the membrane bioreactor operation are considered partially optimized or not optimized, the following steps are recommended for consideration in developing and implementing an action plan.

- Performance targets should be clearly defined for the system and should include effluent discharge requirements, operating targets in accordance with the total sludge mass control program, and the membrane-specific performance (i.e., flux rate and transmembrane pressure).

- A sampling and testing program should be developed that allows for tracking of performance relative to the defined targets. Daily sampling and testing is recommended. All data including relevant performance targets should be plotted on trend charts that are updated and reviewed daily.
- The membrane bioreactor is a variation on the activated sludge process and is perfectly suitable for application of the total sludge mass control program. This is a central component for the optimization of any biological treatment system. Further details have been provided in the suspended growth section of this guide.
- The plant control system should continuously monitor and log performance data for the membrane permeability. This is commonly monitored by measuring the transmembrane pressure and is a critical performance parameter that prompts operating staff when membrane cleaning events are required. Tracking of this information is also useful in assessing the effectiveness of membrane cleaning events (i.e., comparing performance data before and after the cleaning). The goal of membrane cleaning is to restore the permeability as close as possible to the pre-fouled operating condition. Membrane cleaning frequency and protocols should be developed, based on the recommendations of the membrane manufacturer.
- It is recommended that all membrane plants include provision for foam control. The relatively high OLR and, in some systems, high rate of membrane scour during aeration can lead to foaming issues. It is common for membrane plants to include an anti-foam chemical metering system to control foaming issues on an as-needed basis.
- SOPs should be developed for the membrane treatment process that includes an emergency response plan.

Performance Limiting Factors

Table 4-7 summarizes factors related to advanced tertiary treatment that may be limiting performance of the plant. Check whether these factors are Optimized and Documented, Partially Optimized, or Not Optimized. Factors identified as Partially Optimized or Not Optimized will be prioritized in chapter 8, Identification and Prioritization of Performance Limiting Factors, where an action plan can be developed and implemented that allows for optimization of these parameters.

General Optimization Guidelines for Processes Not Specifically Identified in this Guide

The purpose of this guide is to identify methods for optimizing the performance of WWTPs. An almost limitless combination exists of the types and configurations of treatment technologies used at wastewater treatment facilities. This guide has focused on the most common types of processes typically found in the

Table 4-7. Performance limiting factor summary: advanced/tertiary treatment

Self-assessment category	Questions for gauging optimization status	Optimization status		
		Optimized and Documented	Partially Optimized	Not Optimized
Tertiary filtration	Have performance targets been defined for the tertiary filtration process?			
	Has a sampling and testing program been implemented for the tertiary filtration process that allows actual performance to be tracked and compared to the defined targets?			
	If chemical phosphorus removal is included at the plant, has the chemical coagulant dosage been optimized for use with the tertiary filter?			
	Is there a provision for continuous operation during filter cleaning and/or backwashing operation?			
	Have backwash initiation and return-to-service parameters been defined?			
	Does the control system have the ability to automatically initiate a backwash?			
	Has a regular maintenance and inspection program been developed for the tertiary filters?			
	Have SOPs been developed for the tertiary filter that include an emergency response plan?			
Membrane bioreactor process	Have performance targets been defined for the process?			
	Has a sampling and testing program been developed for the membrane process that allows tracking of actual system performance compared with the defined targets?			
	Has the total sludge mass control program been adapted and applied for the membrane bioreactor process?			
	Does the plant control system actively monitor the membrane permeability?			
	Does the system include provision for restoring the permeability of the membranes when they have become fouled?			
	If chemical cleaning is used, has a cleaning frequency and protocol been established?			
	Does plant have an effective method of controlling foam events?			
	Have SOPs been developed for the process that include an emergency response plan?			

industry, with particular focus on processes having the greatest potential impact on effluent quality. Plant operations staff may find themselves searching this guide for assistance on optimizing their facility only to discover that their specific process or technology is not included. This should not discourage operating staff from completing the program and striving for optimal performance. Operators are encouraged to think about their specific process or technology and derive ways the process could be operated more efficiently and effectively. Some common themes in this guide are central to the program regardless of the type of technology. These themes and concepts can be adopted and applied to many other treatment processes or technologies with similar success. In general, the process of achieving optimization can be accomplished by completing the following steps:

- Identify the key operation status questions that would apply to the specific process being optimized.
- Complete a thorough and critical review of all aspects of the process operation, including maintenance.
- Document the results in a table identifying whether each aspect of the operation is fully optimized, partially optimized, or not optimized.
- Develop an action plan that responds to the deficiencies uncovered in the review. The action plan should include specific measures that will be undertaken or implemented to address the deficiencies.

To provide some general guidance for processes not included in this guide, the following list of questions may be used. Operators are encouraged to modify or supplement this list as needed to fully review the operation of their specific process.

- Have performance targets been developed for the process? Existing design documentation or published literature may assist with defining an appropriate set of design targets.
- Has a sampling and testing program been put into place that allows for regular monitoring of the process performance relative to the stated targets? In general, this guide encourages operators to implement a daily testing schedule for key parameters. The exact parameters will depend on the nature of the process. The following general sampling and testing guidelines should be observed:
  - Sample and test key concentrations at both the inlet and outlet of each unit process. This allows calculation and tracking of removal efficiencies within each unit process as well as the entire facility.
  - Test for solids concentrations along with volumes or flow rates to allow calculation of masses or mass rates. This could apply to aeration

tanks, clarifiers, waste sludge streams, recycle sludge streams, and the final effluent, as examples.

- If the process is biological in nature, can the total sludge mass control concept introduced in this guide be applied? The mass control program is quite flexible and can be adopted to most biological treatment processes using activated sludge technology. Information provided in the secondary treatment section of this guide may assist with the implementation of this control program. The Partnership for Clean Water is available to provide guidance on how the mass control program could be adapted for biological processes not specifically covered within this guide or the software.
- Have the appropriate trend charts been developed to monitor and track performance over time? Trend charts should include defined performance targets (where applicable), results of the sampling and testing program, and tracking of the key parameters for the total sludge mass control program.
- Have instrumentation and controls been applied, where appropriate, to provide notification of process deviations and provide process safeguards and controls, where applicable.
- Has a regular inspection and maintenance schedule been developed for the process? Design documentation and equipment literature may assist with identifying maintenance and inspection items as well as maintenance frequencies.
- Have SOPs been developed for the process, including an emergency response plan? Procedures should be documented, readily available to the operations team, and reviewed and updated on a regular basis. Training on SOPs should be provided to all pertinent plant staff. Emergency response plans should include response actions for safety related emergency events (e.g., fire, chemical spill, etc.) as well as process emergencies that may impact the environment (e.g., release of untreated wastewater, foaming, etc.).

SOLIDS TREATMENT PROCESSES

The solids treatment train at a wastewater facility refers to the handling and processing of waste sludge material. Generally, waste sludge is comprised of solids removed from the primary clarifiers, excess biomass from the secondary treatment process, and any solids generated as a result of chemical precipitation in a tertiary treatment step (typically this is generated as a result of chemical phosphorus precipitation). Solids treatment is wide ranging and can include many diverse technologies. The main focus of this section is on the potential impact that solids handling systems can have on the liquid treatment train. As such, this is not intended to be an exhaustive resource on the optimization of all solids handling technologies.

Understanding

The solids treatment train can be considered as a completely separate treatment process from the liquid treatment train. Both processes typically try to achieve completely different goals. The liquid treatment train is generally treating liquid wastewater in preparation for discharge to a receiving water body, whereas the solids treatment train typically treats waste sludge for disposal either in a landfill, in a compost facility, or as a land-applied fertilizer. The connection between the two is any recycle water that is returned from the solids treatment process back to the wastewater treatment process. Typically, recycle flow is generated as a result of dewatering or thickening operations within the solids treatment process. Any soluble contaminants in the waste sludge, such as BOD_5, TP, TKN, etc., would not be captured in the thickened solids. These contaminants pass through the sludge processing equipment into the recycle wastewater stream and are returned to the treatment plant. Common sources of recycle wastewater from solids processing operations could include the following technologies:

- Decant from gravity thickeners, DAF units, drum thickeners, etc.
- Centrate from centrifuge operations.
- Filtrate from belt filter presses, belt thickeners, rotary presses, etc.
- Wash water from equipment cleaning or backflushing operations.
- Excess liquid polymer as a result of polymer overdosing.

All of the above sources of recycle wastewater can have detrimental impacts on the liquid treatment process if not properly controlled. Each operation can introduce new materials not found in typical wastewater (e.g., excess polymer), and can represent increased hydraulic and organic loading. Another example is sludge dewatering. The goal of sludge dewatering is to achieve the highest level of dryness in the sludge cake as possible. Dewatering attempts to generate the maximum volume of water from the waste sludge, which is returned to the liquid process. Depending on the quantity and characteristics of the sludge being processed, this can represent a significant load on the liquid treatment train.

If a WWTP is functioning as a regional solids handling facility, an additional consideration is the processing of sludge from multiple sources. At some communities with several treatment plants, not all facilities will have on-site solids processing capability. It may be that one plant is designated to handle the waste sludge from multiple facilities. Some facilities may also serve as a central septage receiving station for septic haulers. In both of these examples, the plant that receives this material is accepting a significant additional sidestream load that can exacerbate the hydraulic and contaminant loading to the wastewater process. Also, depending on the source of the external liquid and solid material, there is a higher possibility of introducing new contaminants such as metals or nitrification inhibiting substances into the liquid treatment process.

Status

Similar to the wastewater entering the treatment plant from the collection system, the operator may have little control over the quantity and composition of the waste sludge being handled by the solids treatment process. This is particularly true if the facility is a central receiving station for outside sources of sludge and/or septage. The operator, however, needs to manage the recycle flow generated by the solids processing operation and ensure that it does not negatively impact the liquid treatment process. The most important step the operator can take is to diligently monitor the quantity and composition of the recycle flow. Having a good understanding of the quantity and composition will enable the operator to make the necessary adjustments within the liquid treatment process. It may also be an indication of problems or inefficiencies in the solids treatment process.

Some key considerations for the operator with respect to the solids handling process include:

- Some dewatering equipment requires a start-up operational period of several minutes to dial in the polymer dosage and other key settings before a good sludge cake is produced.
  - Often the operator needs to collect samples to visualize the performance of the system during this period and re-adjust the polymer dose or other settings, such as sludge feed rate, to achieve optimal performance.
  - Centrifuges are a prime example of this type of start-up operation that require the performance to be dialed in. Centrifuges need time and adjustment of the polymer settings to generate the initial "plug" of sludge cake. During this initial adjustment period, the equipment typically produces a product that is too thin and is disposed to the liquid train.
  - It may be more prudent to direct this material back to the sludge holding tank where it can be re-fed to the dewatering equipment. Returning the thin polymerized sludge to the plant headworks can result in unnecessary organic loading on the liquid train as well as potentially an excess of unused polymer.
- If the liquid wastewater process is required to nitrify, the operator should be aware that potential exists for a significant TKN load in the solids processing recycle flow. This is particularly true if outside sludge and septage sources are brought to the facility for dewatering. The operator will need to account for the additional TKN and may need to make adjustments to

the target SRT and sludge mass in the secondary liquid treatment process to handle the additional load.

- The operator should have a good awareness of the typical plant diurnal flow variations as there may be opportunity to use these variations as an advantage. For example, it may be beneficial to run the solids handling process during periods of normally low wastewater flow to the plant to reduce the instantaneous hydraulic loading.
- With most dewatering processes, some form of polymer is required to enhance the solids separation.
  - A common misconception is that adding more polymer will improve the dewatering system performance. In some cases, however, overdosing polymer can lead to reduced performance in the dewatering system. Worse still, the excess polymer often passes through the dewatering equipment and is returned to the liquid treatment train.
  - Excess polymer can have detrimental impacts on the liquid process, including undesirable flocculation in the aeration tank and fouling of equipment, such as filters, membranes, and aeration diffusers.
  - Because the characteristics and polymer requirements can change, the operator should complete a jar test before each dewatering operation to confirm the correct polymer dose for the specific batch of sludge being processed.
- If outside sources of sludge and/or septage are delivered to the plant for processing, the operator must have a good understanding of the material's source, quantity, and characteristics.
  - Many plants use a card reader system and flowmeter to record the volume of sludge being delivered by each hauler. The operator should also complete some testing on the various sources of sludge to better understand the characteristics.
  - Of particular concern are any soluble substances that will impact the liquid treatment train such as soluble COD or BOD_5, nutrients (e.g., nitrogen and phosphorus), and inhibitory substances.
  - Better data collection on the source sludge will enable the operator to more effectively determine the source of any substances of concern that may be impacting the liquid treatment train.
- An additional way that excess polymer can be inadvertently directed to the treatment plant headworks is via chemical spills. This can be minimized by including appropriate containment for polymer storage and dosing equipment, and taking care when cleaning up spilled chemical by not washing it down floor drains that lead to the plant headworks.

The following questions are designed to help the operator better understand the optimization status of the plant's solids handling processes. Operators are encouraged to develop and consider additional questions that may be specific to their utility:

- Does the plant have an appropriate method for handling any unsatisfactory polymerized sludge generated during equipment start-up routines? This material should not be directed to the plant headworks if it can be avoided.
- Has a sampling, testing, and monitoring program been developed for the plant that includes recycle wastewater from the solids treatment process? The sampling and testing program should include the following:
  - COD or BOD_5.
  - TKN if the liquid treatment process includes nitrification.
- Has the polymer dosing system been optimized to meet the requirements of the dewatering operation while minimizing excess polymer?
- Does the polymer dosing system operation include the use of jar testing for each new batch of sludge processed?
- Does the sampling and testing program include regular sampling of material from outside sources (i.e., sludge and septage) if applicable?
- Have SOPs been developed for the solids handling process, including an emergency response plan? For the solids treatment process, the emergency response plan should include polymer spill containment and cleanup procedures that prevent spilled polymer from being directed to the liquid treatment process, as well as procedures to address a failure of the plant's solids handling process.

Process Energy Considerations

Depending on the type of solids handling technologies present at a facility, a wide range of process energy considerations may exist. Many of the concepts included in previous sections of this guide would be applicable to some or all of the sludge handling equipment. In general, the following should be considered when attempting to reduce or optimize process energy consumption for the solids handling process.

- Equipment generally operates most efficiently when it is correctly sized and operates as close as possible to optimum efficiency points. Adjusting equipment runtimes and operating frequencies to keep instantaneous speeds or flow rates close to BEPs will help with overall energy efficiency.

- Operating the solids handling equipment during normal periods of average or low flow can improve the energy efficiency of the liquid treatment process by minimizing extreme peak flows for pumping.
- Many solids handling processes include holding tanks or blending tanks that require aeration. Similar to the activated sludge process, operating aeration equipment with variable frequency drives and DO feedback can reduce energy use.
- Reducing the speed or cycling mechanical mixers on and off can help reduce the energy used for mechanical mixing.
- Maintaining sludge storage tank levels as low as possible between dewatering events may reduce the energy demand for aeration and/or mixing systems.
- Proper maintenance of equipment, including regular cleaning and backflushing, will ensure that equipment is able to operate at or close to design operating targets, including throughput volume. Dirty, plugged, or improperly maintained equipment may require longer runtimes to process the same volume, which will lead to excess energy consumption.

An additional area of potential energy optimization exists for facilities that use anaerobic treatment for waste sludge digestion. The anaerobic fermentation process converts organic material (i.e., biodegradable COD) in the waste sludge into methane gas. The methane gas is typically collected in the headspace of the sealed reactor and must be properly handled due to its hazardous nature. With proper equipment in place, it is possible to capture the gas and use it as a renewable energy source to generate heat or electricity. In either case, the use of this material as an energy source offsets plant energy consumption from the electricity grid or through burning of conventional fossil fuels. At some WWTPs, it may be advantageous to install an anaerobic digester for the sole purpose of generating this renewable energy source. Both the quantity and quality of the sludge to be digested will impact the payback period to recover costs to construct an anaerobic digester. Plants considering adding anaerobic treatment or adding biogas utilization equipment are advised to complete a thorough cost-benefit analysis to determine the potential energy savings.

Action

If areas of the solids handling operation are considered Partially Optimized or Not Optimized, the following steps are recommended for implementing an action plan.

- The plant should develop a procedure for proper equipment start-up that includes handling of polymerized sludge that is too thin for disposal with the dewatered solids. Operations staff should consult operation and maintenance manuals and/or the equipment manufacturer for additional guidance on proper equipment start-up routines to minimize the volume of this material.

- A sampling and testing program should be developed for the solids treatment process. To help minimize impacts on the liquid treatment process, the operator should regularly test the wastewater recycle stream as well as regularly test batches of any outside sludge material delivered to the plant.
- The plant should implement a regular jar testing procedure to optimize polymer dosing. Jar testing should be completed before each new batch of sludge is dewatered to ensure that any inconsistencies in the sludge characteristics are accounted for. The goal should be to minimize excess polymer returned to the liquid treatment train.
- SOPs should be developed for the plant that include an emergency response plan and spill containment and cleanup procedures for spilled polymer, as well as a plan to address failure of the plant's solids handling processes.

Performance Limiting Factors

Table 4-8 summarizes factors related to the solids treatment process that may be impacting the performance of the solids treatment train. Check whether these factors are Optimized and Documented, Partially Optimized, or Not Optimized. Factors identified as Partially Optimized or Not Optimized will be prioritized in chapter 8, Identification and Prioritization of Performance Limiting Factors/ Action Plan Development, where an action plan can be developed and implemented that allows for optimization of these parameters.

References

Bowker, R.P.G., Stensel, D.H., Hartmann, G.L., and Smith, J.M. 1987. *Design Manual: Phosphorous Removal*. Cincinnati, OH: U.S. Environment Protection Agency.

Tchobanoglous, G., Stensel, D.H., Tsuchihashi, R., and Burton, F. 2014. *Wastewater Engineering: Treatment and Resource Recovery, 5th Ed.* New York: Metcalf & Eddy Inc., McGraw-Hill.

Water Environment Federation. 2010. *Design of Municipal Wastewater Treatment Plants, 5th Ed., WEF Manual of Practice No.* 8. New York: McGraw-Hill.

Water Environment Federation. 2011. *Nutrient Removal, WEF Manual of Practice No.* 34. New York: McGraw-Hill.

Water Environment Federation. 2011. *Membrane Bioreactors, WEF Manual of Practice No.* 36. New York: McGraw-Hill.

Table 4-8. Performance limiting factor summary: solids treatment processes

Self-assessment category	Questions for gauging optimization status	Optimization status		
		Optimized and Documented	Partially Optimized	Not Optimized
Sludge dewatering	Does the plant have an appropriate method for handling any unsatisfactory polymerized sludge generated during equipment start-up routines?			
	Has a sampling, testing, and monitoring program been developed for the plant that includes recycle wastewater from the solids treatment process?			
	Has the polymer dosing system been optimized to meet the requirements of the dewatering operation while minimizing excess polymer?			
	Does the polymer dosing system operation include the use of jar testing for each new batch of sludge processed?			
	Does the sampling and testing program include regular sampling of material from outside sources (i.e., sludge and septage) if applicable?			
	Have SOPs developed for the solids handling process, including an emergency response plan?			

CHAPTER 5

FACILITY ENERGY OPTIMIZATION

Introduction to Facility Energy Optimization

This chapter explores the multiple facets of facility energy optimization with a focus on building systems. Wastewater treatment plants (WWTPs) are significant consumers of power, both in their treatment processes, as discussed in chapter 4, and in their building systems. Many of the same self-assessment principles apply to both process loads and building systems, but the assessment and optimization of process-related energy consumption is included within the applicable process performance evaluation sections in chapter 4. Chapter 5 pertains to the self-assessment and optimization of the facility's building systems.

The following facility energy optimization topics are covered in this chapter:

- Buildings portfolio assessment
  - The purpose of this assessment is to help facility managers understand the opportunities with which applying effort and funds to reduce energy consumption may achieve the most benefit.
- Energy auditing
  - This section describes strategies for energy auditing as well as typically implemented building system replacements and improvements.
- Re- or retro-commissioning
  - This section provides energy considerations for facilities not yet ready for major renovations but that are due for a tune-up.
- Energy management, measurement, and verification
  - While energy audits and re- or retro-commissioning typically occur infrequently in a building's life span, energy management, measurement, and verification can occur continuously if resources are allotted to this important endeavor.
- Energy codes and standards
- Third-party certification options

There is overlap among the topics presented in this chapter. In the end, they all seek to facilitate energy optimization. Depending on the size and age of the facility; the number of facility managers; and budgets, operations, and maintenance practices, there will be an approach or combination thereof that is more appropriate than another for each WWTP. Thinking about this and establishing a facility optimization action plan is the proposed objective for utilities completing the WWTP self-assessment process.

As described in chapter 1 of this guide, the self-assessment process is optimally completed by engaging a diverse team in the process, including operations staff. This also applies to the facility energy assessment included in this chapter. Although the material included in this chapter may not be as familiar to operations staff as that included in the unit process performance assessment and in other chapters, input and observations from staff at all levels may contribute to the identification of performance limiting factors related to energy optimization, for which an action plan may be developed and implemented. It is recommended that completion of this section of the self-assessment process be conducted similarly to that of any other chapter of this guide. For example, if the self-assessment questions have been discussed in a series of team meetings, continue to hold team meetings for the facility energy assessment material, including, if possible, utility subject matter experts relating to energy efficiency. As with other chapters of this guide, it can be helpful to fully review the chapter prior to beginning the self-assessment process and to compile basic information relating to facility construction and energy consumption for review and discussion by the self-assessment team.

Preliminary Assessment: How Well Are Your Buildings Doing?

Understanding

For facilities comprising several buildings of different vintages, understanding how each is performing will provide facility managers with useful data to develop an action plan outlining where focusing effort and funds may achieve the most benefit. In some facilities, this information may already be available and compiled either as part of operations and maintenance programs or in energy monitoring–measurement–verification programs. Several portfolio management programs and varieties of software are available to assist with this. One of them, Portfolio Manager, is discussed in subsequent sections. Compiling and assessing this information can be considered to be a performance assessment for facility energy consumption, similar to the performance assessment that was completed for WWTP processes included in chapter 2. This process allows utility staff to identify the current status of the facility's energy consumption.

At a minimum, compiling a spreadsheet of what information is available and known by facility managers can provide a useful description of the current situation. It is important to note that building systems, such as heating, cooling, ventilation, and lighting, may represent on the order of 30 percent of a WWTP's total energy consumption. However, process loads can completely dwarf building system loads in certain key process load buildings. The ratio is very site-specific and will vary widely from one facility to the next; it will be smaller in climates with low heating and cooling loads or where there is no administration building.

Nevertheless, significant amounts of energy consumption and costs are associated with building systems, and it often makes good financial sense to implement improvements in this area. Implementing measures to reduce building system energy consumption can often be simpler than trying to reduce process energy consumption as changes to building systems do not affect the quality of the treated wastewater effluent.

Building Energy Performance Assessment

How can building energy efficiency performance be quantified such that the most impactful opportunities for improvements may be identified? The industry uses benchmarking metrics, such as Energy Use Intensity (EUI). EUI is expressed in units of kBtu/SF-yr or GJ/m^2-yr to quantify energy use normalized to building size. Although a variety of units and methods may be used to quantify energy consumption, the important thing is to choose one system and remain consistent for the purposes of the current performance assessment and year-to-year comparisons.

The objective of this exercise is to allow buildings of varying sizes, energy sources, and energy suppliers to be compared with one another. Within buildings of the same type and occupancy, comparing the EUI will allow staff to determine which building consumes the most energy per building area per year. However, in buildings in which process loads dominate energy consumption, and in particular when energy metering does not separate usages, additional effort and knowledge will be required to understand actual trends. The advantage of determining the EUI is that it is a comparison of energy consumed per building area per year rather than a simple cost comparison.

The US Environmental Protection Agency's (USEPA's) Energy Star® program is a voluntary program, founded in 1992, that provides a variety of free tools and resources to help organizations and individuals quantify energy usage and improve energy efficiency, reducing energy costs and greenhouse gas emissions. One tool provided through the Energy Star program is the Energy Star Portfolio Manager®. The Portfolio Manager is an energy measurement and tracking tool for commercial buildings. The Portfolio Manager has a tool specific to WWTPs that enables calculation of EUIs, including process data. The Energy Star program is described in greater detail later in this chapter.

Natural Resources Canada (NRCan) also offers a free and easy-to-use website, which allows users to enter a building's annual energy consumption to obtain the EUI. The site may be accessed at oee.nrcan.gc.ca/commercial/technical-info/tools/gigajoule.cfm?attr=20.

The tool requires entry of the building's utility bills for the past year, differentiating entries based on energy source (electricity, oil, natural gas, steam) along with the building's area to calculate the building's estimated EUI. This exercise may be repeated for multiple buildings or multiple facilities, and the results may be recorded into a utility-created spreadsheet to assist in tracking changes in EUI that may occur over time. An example of NRCan's Gigajoule and Energy Intensity Calculator, along with a sample building energy tracking spreadsheet, are displayed in Figure 5-1 and Table 5-1.

Gigajoule and Energy Intensity Calculator

This tool helps you convert various energy measurements into gigajoules and then determine your building's overall energy intensity. Read a definition of gigajoule (GJ).

The values for steam and wood are for estimation purposes only. Steam energy content varies with pressure and feed water temperature, and the calculator uses the industry accepted estimate used when exact steam conditions are not known (1000 Btu/lb or 2.33 MJ/kg). The heating value of wood varies based on density, moisture, and ash content.

Do not use commas; a comma (,) acts as a decimal point (.)

Gigajoule and Energy Intensity Calculator

Electricity	193719	kWh ×	0.0036	697.3884	GJ
Oil #2 (light)		litres ×	0.03880	0	GJ
Oil #6 (heavy)		litres ×	0.04250	0	GJ
Propane		litres ×	0.02531	0	GJ
Natural Gas #1		m^3 ×	0.03826	0	GJ
Natural Gas #2	4989	GJ ×	1	4989	GJ
Steam		kg ×	0.00233	0	GJ
Wood		kg ×	0.018	0	GJ
Other		GJ			
Total	198708			5686.3884	GJ
Area in square feet		ft^2 ÷	10.76	0	m^2
OR					
Area in square meters	5500	m^2		5500	m^2
Energy Intensity	1.0338888	GJ/m^2			

Calculate Reset

Source: Natural Resources Canada oee.nrcan.gc.ca/commercial/technical-info/tools/gigajoule.cfm?attr=20

Figure 5-1. Screenshot of the Natural Resources Canada Gigajoule and Energy Intensity Calculator-online tool

Table 5-1. Example of a spreadsheet that may be used to compile EUI results

Property	EUI (GJ/m^2-yr)
Building A	1.03
Building B	0.4
Building C	1.16

Establishing Priorities

It is suggested that the building(s) with the highest EUI become the highest priorities for energy audits and renovation projects as they may have the greatest potential for reductions in energy consumption and its associated costs. In Table 5-1, Building C would receive the highest priority for further energy optimization activities on the basis of it having a higher EUI than the other buildings located at the WWTP site. This EUI priority list can be compared with other planning tools, such as preventative maintenance planning, in order to create and implement a facility optimization action plan.

Status

The following questions are designed to help plant staff better understand the status of the WWTP's building energy performance. Staff are also encouraged to develop and consider other questions that may be specific to their utility:

- Does the wastewater treatment facility currently quantify energy consumption using energy monitoring, measurement, and verification programs and portfolio management tools that provide this information?
- Has the total number of buildings on the WWTP site been quantified?
- Have all of the different fuel sources used by the facility been quantified?
- Is process energy consumption metered separately from the energy consumed by building systems?
- Is the appropriate information (e.g., utility bills) currently compiled in a format that allows utility staff to complete this exercise?

Action

If areas of the building energy performance assessment are considered Partially Optimized or Not Optimized, the following steps are recommended for developing and implementing an action plan:

- If the wastewater treatment facility has no formal process to quantify energy consumption, consider implementing such a process using some of the readily available energy quantification tools, such as those available through Energy Star, NRCan, or other organizations. Regularly tracking energy consumption can help utilities identify trends and prioritize opportunities for potential future improvements.

- If utility billing information is not available in a format that allows staff to readily quantify energy consumption through measures such as EUI, work with utility accounting staff or with the plant's local energy provider to determine if this information can be made available in the required format. Utility staff may also work with their local energy provider to determine the sources of energy used to supply the WWTP if this information is unknown.
- In cases with incomplete knowledge of the WWTP site, a site plan should be referenced or developed that identifies all buildings to be considered in the energy assessment.
- If process energy consumption is not currently reported separately from building energy consumption, determine if these quantities may be separately reported for more detailed site analysis.

In cases in which the information available is sufficient and the building performance assessment was determined to be a useful exercise, plant staff should proceed with rating the buildings' portfolio. The conclusions can be used to help facility managers understand where applying effort and funds toward facility energy optimization may achieve the most benefit.

Energy Audits

Understanding

Energy audits are an evaluation of facility building systems and are designed to determine where, when, why, and how energy is being used as well as identify and develop opportunities to improve energy efficiency. Implementation of the recommendations provided as a result of completing an energy audit can decrease energy costs and reduce greenhouse gas emissions.

Energy audits typically focus on the building systems that are common to most buildings, including the building envelope; heating, ventilation, and air conditioning (HVAC) systems; lighting; domestic hot water; and miscellaneous electrical loads (computers and such). This chapter focuses on these typical building systems. Energy audits for buildings containing process loads may or may not address these loads aside from perhaps identifying the portion of energy they consume. When seeking to hire consultants to conduct energy audits, the project's scope-of-work definition becomes key to ensuring whether this aspect is included or not. For additional information, refer to the section of this chapter regarding re- or retro-commissioning as energy auditing can, if specified, include aspects of this exercise.

Energy audits can be conducted by generalists, when only one person does the work, or they can be conducted with involvement by a professional from every discipline (typically architectural, mechanical, and electrical). Further, the reports obtained from a professionally conducted energy audit can vary greatly in their depth. A clear scope-of-work definition is required to differentiate between a walk-through audit and an in-depth study. Either type of energy audit may be sufficient based on the plant's specific objectives. Many jurisdictions and organizations have funding available to help fund a portion of the costs associated with energy audit activities and implementation of energy efficiency measures. Utilities are encouraged to verify program details and eligibility prior to beginning any work. Failure to proceed in the required steps may result in program ineligibility.

Several associations, government departments, and other organizations publish information, literature, tools, and professional development resources related to energy auditing. Some of these resources are specific to WWTPs, as noted here:

- National Renewable Energy Laboratory (NREL) has published a document titled "Energy Efficiency Strategies for Municipal Wastewater Treatment Facilities," written by J. Daw and K. Hallett of NREL and J. DeWolfe and I. Venner of Malcolm Pirnie. The document contains information specific to WWTP energy auditing and may be accessed using the following link: nrel.gov/docs/fy12osti/53341.pdf. The document provides a typical breakdown of WWTP energy usage. The aeration process is typically the largest energy consumer for most WWTPs, as indicated in the document, as well as is discussed in chapter 4 of this guide.
- The Association of Energy Engineers offers training, resources, and many different professional certifications. Certified energy auditor (CEA) is one of this organization's commonly known designations. Additional information about CEA certification may be accessed using the following link: aeecenter.org/i4a/pages/index.cfm?pageID=4552
- The Government of Canada, in particular NRCan, also offers an important array of tools on energy auditing. One of these tools includes the Energy Savings Toolbox – An Energy Audit Manual and Tool, developed in cooperation with the Canadian Industry Program for Energy Conservation. This resource may be accessed by using the following link: nrcan.gc.ca/sites/www.nrcan.gc.ca/files/oee/pdf/publications/infosource/pub/cipec/energyauditmanualandtool.pdf
- Additional resources are available through the American Society of Heating, Refrigerating and Air-Conditioning Engineers (ASHRAE), a society focused on promoting sustainable technology for the built environment. ASHRAE offers training, books, and templates for energy auditing. Professionals may also earn an ASHRAE certification related to energy assessment, such as building energy assessment professional or building energy modeling professional. The organization's website may be accessed at ashrae.org.

ASHRAE energy audit level definitions are commonly used to define the scope of work when hiring energy audit consultants. Basic definitions of these levels are provided here for WWTP staff potentially interested in learning more about performing a facility energy audit. The three levels, as defined by ASHRAE, are as follows:

- ASHRAE level I: Walk-through assessment audits are a basic walk-through and summary report that may be used as a starting point for further study.
- ASHRAE level II: Energy survey and analysis audits are perhaps the most common and include a detailed building survey and energy analysis with energy modeling.
- ASHRAE level III: Detailed analysis–modeling audits are in-depth, longer-term studies, providing more precise forecasting of energy savings potential and construction cost estimating.

Additional information about ASHRAE's energy audit procedures may be accessed in *Procedures for Commercial Building Energy Audits, Second Edition*. Additional information about this book, along with previews of typical energy audit templates, may be accessed using the following link: ashrae.org/resources--publications/bookstore/procedures-for-commercial-building-energy-audits

The following sections provide a more in-depth description of the areas covered in a typical energy audit process.

Energy Audit: Site and Data Review

Energy auditors will typically begin their work by reviewing construction documents to understand the building systems. They will then review the buildings from top to bottom and have discussions with project stakeholders. Staff knowledgeable in the operations and maintenance of the building are a valuable asset during these site reviews, providing insight into system operating schedules and condition as well as access to mechanical and electrical spaces. A meeting covering topics such as the motivation for the energy audit, the expected outcomes from the project, and any special considerations in regards to upgrades or budget implications and restrictions is recommended. The auditor will use this information to make more relevant recommendations in the report.

Energy Audit: Energy Modeling and Load Profile

Energy auditors have a variety of tools available with which to model energy consumption. These may include simple tools, such as calculation spreadsheets, or full building energy models, such as those made with eQUEST, an energy modeling tool developed by an industry consultant in collaboration with Lawrence

Berkeley National Laboratory and the US Department of Energy. The energy model used by the energy auditor will be calibrated to represent the energy usage of the facility. Sometimes, multiple tools are used and the results combined to represent situations that don't fit the confines of one particular software tool. What is most important is that the cause and effect implications of actions taken to improve energy efficiency be understood and included in the analysis; for instance, a savings in lighting energy can create a slight increase in heating requirements as energy-efficient lights, such as LEDs, emit less heat than incandescent or fluorescent light bulbs. Conversely, the same change in lighting energy can lower cooling costs. The objective of creating the energy model is to simulate all the buildings' current annual energy consumption profile using historical weather data, historical energy data, and information obtained during the site review.

Energy end use distributions (EUDs) and energy load profiles (ELPs) will typically be created from the energy model data. The EUDs provide a breakdown of the buildings' annual energy consumption with numerical and percentile values detailing the amount of energy consumed by each of the following building systems:

- Heating (through terminal devices and ventilation)
- Ventilation
- Air conditioning
- Lighting
- Domestic hot water
- Miscellaneous electrical loads
- Process loads

Breakdowns are then compared with benchmark data consisting of energy consumption for similar buildings operating in similar environments. The ELPs are graphical representations of the facility's monthly energy consumption over a one-year period, otherwise known as the base year.

Energy Audit: Building Systems, Typical Energy Efficiency Measures

Building Envelope

Building envelope considerations include construction characteristics of doors, windows, floors, walls, and roofs. Insulation levels, thermal bridges, air leakage, and—in the case of windows—solar heat gain coefficients are all factors to consider when trying to keep the heat in during heating season or keep the heat out during cooling season. In many jurisdictions, standard new construction insulation levels are higher than in the past as high-energy costs, energy codes and guidelines, and concerns for the environment encourage architects and owners to invest in better envelopes up-front. It is important to invest in good building envelopes; even the best HVAC systems cannot make up for a poor building envelope. Building envelopes are of critical importance to achieve energy efficiency. If the building envelope is very poor, even improving the heating and cooling systems may not allow facilities to achieve energy efficiency. Some building envelope considerations are described in greater detail in the following paragraphs.

Well-insulated walls and roofs contribute to less energy consumption by slowing down the rate of heat exchange between the indoors and outdoors. When considering wall and roof construction, it is important to consider thermal bridging and to calculate the overall wall assembly R-value (thermal resistance). It can be quite surprising how much heat is lost through metal studs despite excellent adjacent insulation.

The window-to-wall ratio should be kept below 40 percent as a general guideline. Some codes require users to trade off higher ratios for other energy-efficiency features to make up the difference. Floor-to-ceiling windows are not a requirement to achieve good interior daylighting and can contribute to discomfort for occupants as a result of solar glare and air currents. Air leakage, frame type, and type of operability are important considerations when choosing windows. Triple-pane windows are becoming increasingly popular, particularly in colder areas. They are more expensive but provide a better insulation value. When clarity and visibility are not a priority (clerestory for example), windows can be replaced with products that allow light to come in but offer greater insulation value.

Air tightness is desirable to control air changes. Windows, doors, construction joints, duct dampers, and envelope penetrations need to have functional seals. In existing buildings, it is not always easy to increase insulation levels without major renovations. Some of the easier applications with reasonably quick payback periods include

- adding insulation in the attic,
- replacing old single-pane windows with good double- or triple-pane windows,
- maintaining weather seals around doors and windows to reduce air leakage,
- removing windows and replacing them with a wall assembly where space usage has changed and does not benefit from a window, and
- adding insulation during planned roof and siding renovations.

Improving wall insulation on buildings with brick and stone exteriors typically involves removing interior wall finishes and installing insulation from the interior of the building. This can often result in losing several inches of conditioned space along the perimeter. Sometimes, there are empty cavities that can be filled with insulation, avoiding the loss of conditioned space.

Prior to implementing such renovations, consider potential unintended consequences, such as locking migrating moisture inside wall or roof assemblies—which can lead to mold and building deterioration—or exposing existing items to conditions more extreme than they are suited for, again leading to deterioration or damage.

Architects or building science professionals should be retained to consider the existing conditions and the climate particular to the plant location prior to proceeding. In some cases, it is also necessary to consult with a structural engineer. For example, in areas where snow loads are important, adding insulation can reduce the amount of snow melting throughout winter due to heat loss. This can lead to thicker snow accumulation on the roof than what the structure was designed to handle, risking structure failure and obviously needing to be avoided.

To choose better building envelope products—a potential resulting action of completing an energy audit—it is also important to understand related terminology. Conduction heat transfer rate is inversely proportional to thermal resistance, in other words, the higher the R-value (IP units: hr-ft^2-F/Btu) or RSI (SI units: K-m^2/W), the lower the heat transfer rate. Note that thermal resistance of a wall assembly (R-value or RSI) is not simply the value of the insulation layer; it must consider all layers, including thermal bridging, such as wall studs, which can reduce the overall thermal resistance considerably.

Window performance is typically rated by U-factors (thermal transmittance) U = 1/RSI or 1/R-value (be aware of the units) and solar heat gain coefficient (SHGC), which is important in determining solar heat gain in cooling calculations. Opposite to R-value, lower is better concerning U-factors and SHGC. However, if the particular building and/or region does not have or require cooling, a higher SHGC is acceptable, and the U-factor becomes the primary focus.

Heating, Ventilation, Air Conditioning, and Mechanical Systems

HVAC systems are often among the largest energy-consuming systems in a building, surpassed only by plant process loads. Climate, occupancy schedule and type, and quality of the building envelope are the factors external to the systems themselves that define the design loads and affect how much output is required from HVAC systems in meeting the requirements of the building. Choice of HVAC system types, quality of design and control sequences, and condition of the equipment consequently affect how much energy is input into the HVAC system in order to output the effort required to meet the requirements of the building. Manufacturers are constantly innovating, and energy codes are pushing the market toward higher efficiency systems.

A utility's choice of the best systems for a building depends on location, climate, application, stringency of comfort requirements, availability of fuels, availability of space to install mechanical systems, utility costs, budgetary restrictions, and local jurisdiction codes and bylaws. Retaining mechanical engineers to conduct such evaluations is recommended. The evaluation may also involve electrical engineers and structural engineers as changing mechanical systems may change power requirements and may require a structural review (of the floor or roof-bearing capacity) prior to proceeding.

HVAC systems are part of the broader mechanical systems category that includes, among other items, domestic hot water and fire protection. These systems may also have associated pumps and energy consumption considerations. In the case of domestic hot water, heating is sometimes provided by the same systems providing HVAC services. The plant's quantity of hot water usage can vary greatly by building occupancy type, and it can be fueled by electricity, natural gas, propane, oil, or even heat pumps. Sometimes, waste system or process heat can be recovered to reduce or offset energy requirements for heating or preheating of domestic hot water.

Energy efficiency may be improved in mechanical systems in various ways by following these general guidelines:

- Turning systems off when not required
- Choosing energy efficient equipment for HVAC
- Ensuring equipment is properly sized for the task
- Allowing equipment to operate in its most efficient mode
- Considering efficiency of motors for fans, pumps, and variable speed motors
- Avoiding simultaneous heating and cooling and other waste
- Using "free cooling" when available
- Recovering energy when available
- Using pipe and duct insulation and motorized dampers on louvers
- Applying programmable thermostats and buildings controls
- Installing low-flow plumbing fixtures to reduce hot water usage

Examples of energy-efficient HVAC equipment are

- condensing natural gas boilers,
- hydronic or radiant heating,
- heat recovery ventilators,
- fans and pumps with variable speeds,

- air destratification fans, and
- heat pumps and geothermal systems.

Examples of renewable energy mechanical systems are

- biomass boilers (wood chips or pellets) and
- solar thermal domestic hot water.

WWTPs may also have the option of recovering energy generated by the anaerobic digester system to provide power to the greater facility, potentially resulting in reduced energy costs.

In order to choose energy-efficient equipment, a familiarity with industry terminology is necessary. Common energy units are the kilowatt-hour (kWh) and British thermal unit (Btu). They represent the amount of energy consumed or the power over a set period of time. Common power units are the Watt (W), kilowatt (kW), British thermal unit per hour (Btu/h), and horsepower (hp).

When discussing heat pumps (and other refrigeration equipment), several terms are commonly used. The *coefficient of performance*, or COP, is a ratio between the useful cooling capacity and the equivalent power input (both in kW). A COP of three indicates that the refrigeration cycle can remove three times as much from the building than it is consuming in power. The *energy efficiency ratio*, or EER, is the same ratio but in units that are more intuitive to some consumers, using Btu/h for the cooling capacity and W for the power input. The seasonal energy efficiency ratio (SEER) can be used to represent conditions over a season. In heating mode, the COP becomes the COP_h (with the useful heat output as numerator) and the heating seasonal performance factor to provide indication of performance over a season rather than at a single standard condition.

With all of these parameters, a higher value is better. Verify the heat pump's efficiency for the local climate data prior to purchasing as these ratings are for standard data only. At low temperatures, heat pumps will need to be backed up by electric heaters in some climates. Boiler efficiency, expressed as a percentage, is the ratio of heat output to heat input, representing the heat capacity of amount of fuel burned versus the useful heat produced. Replacing old boilers with new, more efficient ones often leads to significant savings. Natural gas condensing boilers typically achieve efficiencies over 90 percent.

Ventilation can be building occupancy– or process-driven. ASHRAE 62 is a common code requirement for building ventilation rates for common occupancy types. Industrial processes typically require exhaust ventilation rates above ASHRAE 62—often to protect occupants from harmful (sometimes deadly) concentrations of pollutants. In hazardous classification areas such as those defined by the National Fire Protection Agency (NFPA), certain minimum ventilation rates may be required to avoid potentially explosive environments. Lowering ventilation rates can thus be dangerous; it is advised to proceed with caution and consult with experts if considering making modifications in this area.

Heat and energy-recovery ventilation allow for recovery of waste energy in airflow streams, and the efficiency typically relates to the amount of heat (sensible) or energy transfer (sensible and latent or total). When choosing a system, consider all aspects, such as additional fan power, types of defrost systems and implications, ducting, type of heat or energy exchanger, tolerable cross leakage, and whether latent energy recovery is desirable.

Lighting

Lighting can represent a very large portion, even the majority, of a building's energy costs if there are low heating and cooling needs and low process loads. In many existing buildings, making an effort to improve lighting systems is worthwhile. An office building may see on the order of a quarter of its energy costs associated with lighting. The industry has seen great strides in energy efficiency with the development of affordable light-emitting diode (LED) lighting options as well as sophisticated daylighting controls. Many new buildings are installing only LED lighting and foregoing fluorescent lights altogether (which has the added benefit of reducing mercury content of the lighting equipment—a potent pollutant). Older buildings, using their original incandescent and industrial lighting, have the most to gain by upgrading lighting systems with striking efficiency improvements available at reasonable costs. Even newer buildings may be able to conduct a lighting upgrade, achieving payback periods of less than 10 years.

LED lighting also provides savings associated with reduced maintenance costs. For example, many LED fixtures are rated to maintain 70 percent of their output for 50,000 hours or 24 years on the basis of lights being operated 40 hours per week. This reduces maintenance costs associated with light bulb replacement, including labor costs.

Energy savings associated with lighting can be obtained by considering some of the following actions:

- Ensuring lights are off when not in use. In some spaces, it can make sense to install occupancy sensors, daylight sensors, or automatically controlled lighting schedules.
- Ensuring that lighting intensity is appropriate for the occupancy type (do not over-illuminate).
- Replacing light fixtures with more efficient models: LEDs are now commonly available, have great longevity, and costs have come down.

In order to choose an energy-efficient light source, compare W consumed and lumens (lm) emitted. Lighting options with lower W and higher lm are most desirable. Efficacy is a term used for evaluating lighting sources and is in units of lm/W. Green building rating systems, such as Leadership in Energy and Environmental Design (LEED), also encourage selection of higher quality lighting, defining additional parameters with which to compare products.

Miscellaneous Loads

Miscellaneous loads are energy expenditures particular to building occupancy type and usage. An example of a miscellaneous load is computers and monitors in office buildings. Energy usage can be reduced by turning equipment off when not in use and replacing obsolete equipment with more efficient equipment. Programs, such as Energy Star and EnerGuide, discussed in other sections of this chapter, can help utilities choose efficient equipment and operate the equipment in the most efficient manner possible.

Demand Charges and Off-Peak Tariffs

Understanding demand charges and utility tariff schedules is an important consideration toward controlling energy costs. While this may not reduce the plant's overall energy usage, it can considerably reduce the utility's energy costs. Reducing demand charges can yield important savings for WWTPs as process equipment have large motors with significant power consumption. By flattening pump demands during peak periods, for example, the utility may be able to realize an energy cost savings associated with strategic pump operation. Shifting loads to off-peak periods can be an important consideration in jurisdictions with lower off-peak tariffs.

Status

The following questions are designed to help the plant staff better understand the utility's status with respect to the completion of an energy audit. Staff are also encouraged to develop and consider other questions that may be specific to their utility:

- Has the utility completed an energy audit for either a single building or the entire facility?
- Has the utility defined the needs and scope of work related to energy audits and efficiency upgrades?
- Is the utility familiar with local codes and bylaws that affect implementation of energy-efficiency upgrades at the facility?
- Are any major building renovations planned at the facility? If so, are there energy-efficient upgrades that would be beneficial to implement during these planned renovations?
- Are utility staff aware of programs applicable to the building or project that would provide financial incentives to improve energy efficiency? If

so, are utility staff aware of the rules and procedures required for eligibility to receive financial incentives?

- Have subject matter experts been identified to assist the utility in moving project plans forward in this area?

Action

If areas of the energy audit assessment are considered Partially Optimized or Not Optimized, the following steps are recommended for developing and implementing an action plan.

- If an energy audit has not been performed at the site, consider developing plans to complete an energy audit for a single building or the entire site.
  - Use the energy performance assessment data and other information to help prioritize site facilities for completion of an energy audit.
  - When planning for completion of an energy audit, it is important to define the scope and objectives of the audit prior to initiating the process. It is essential to begin with the end in mind.
  - When major renovations are planned, evidence may show that the facility is a high-energy user, and budget is available, an entire facility energy audit is recommended in order to identify energy-efficiency upgrades that may be completed in conjunction with the facility renovations. It can be most cost-effective to complete energy upgrades concurrently with other physical building renovations rather than as a separate project.
  - If there is no budget for or intent of implementing renovations and upgrades, consider a lower cost walk-through assessment-type energy audit. Another option is to consider energy management or recommissioning exercises with a focus on the "quick hitters" and "low-hanging fruit" rather than system replacements.
  - Consider completing an energy audit for a single building—perhaps the one that appears to have the highest potential for improvement—as a starting point.
- Utility staff and their contractors or consultants should be aware of all applicable energy codes and regulations in their area. If newer energy codes are applicable in the plant's area, the audit can include measures required to make the buildings code-compliant. In practice, some of these recommendations may not make financial sense, but it is an insightful exercise.
- When completing renovations, utility staff should research and be aware of all applicable financial incentives from project initiation. Ensure that

the facility plan meets the requirements of any applicable financial incentives; the rules can typically be fairly rigid.

- Consider identifying subject matter experts and consultants who can potentially be retained to assist with the energy auditing process.
  - If selecting local expertise proves difficult, a public request for proposals is suggested.
  - Defining a clear scope of work will help ensure that the energy audit yields useful results and that contractor bids are as accurate as possible.
  - Hiring consultants to complete an entire facility energy audit at once may produce a more competitive bid than an energy audit for a single building.
- A more detailed list of ideas for energy efficiency measures can be found on the ASHRAE website, using the following link: xp20.ashrae.org/PCBEA/Files/EEMs-to-Consider-2011-09-15.pdf

Re- or Retro-Commissioning

Understanding

Re- or retro-commissioning is an exercise similar to energy auditing that provides recommendations that improve performance in commissioning existing buildings. This process can be called recommissioning when the buildings were commissioned prior to the first occupancy or retro-commissioning when the buildings were not. Thus, a retro-commissioning exercise is often more involved than recommissioning. For the purpose of this text, the two terms will not be further differentiated, and RCx, a common acronym, will be used to represent the concept.

The following definition, provided by a California utility commercial buildings RCx rebate program, is concise:

> Retrocommissioning (RCx) is a systematic process for identifying less-than-optimal performance in your facility's equipment, lighting and control systems and making the necessary adjustments. While retrofitting involves replacing outdated equipment, RCx focuses on improving the efficiency of what's already in place. (source: pge.com)

In a typical RCx scope, more labor is spent than during an energy audit because the intent is to fine-tune existing systems. This labor uncovers a category of recommendations that often have very little cost associated and can result in significant energy savings. It will also uncover recommendations that can improve outcomes, such as comfort of occupants.

Instrumentation used by RCx agents can include ambient temperature and humidity measurements, airflow and hydronic system flow measurements, pressure and fluid temperature measurements, and various types of electrical measurements ranging from simple current clamps to the quality of power supply. Data-logging equipment may be installed in the facilities and used to collect data over longer periods of time. In comparison with a traditional energy audit, an RCx provider typically has mechanical and electrical staff and uses data collection instrumentation whereas an energy audit would also include architectural input for the building envelope and may not include data collection instrumentation. Energy audits typically are a higher level view of how efficient the systems are, and though there may be overlap between the two exercises, the energy audit recommendations typically include big ticket retrofit items that are more convenient to implement during large renovations.

Scope variation is common; thus, when hiring contractors or consultants to perform RCx and/or energy audits, it is important to understand what their service includes. A utility can choose to complete both facets (RCx and energy audit) in one combined exercise or perform a single exercise at a time.

A variety of RCx resources are available online, including online training courses. Key Energy Star and NRCan publications are highlighted here:

- Chapter 5 of the Energy Star program's *Building Upgrade Manual* focuses specifically on RCx and includes details about the RCx process from project planning and execution to more specific tune-up opportunities. The document may be accessed using the following link: energystar.gov/buildings/tools-and-resources/energy-star-building-upgrade-manual-chapter-5-retrocommissioning
- The NRCan website also has valuable information, including an RCx guide for building owners and managers that may be accessed using the following link. This guide presents information about the RCx process, process guidelines, potential cost reduction options, and RCx benefits. nrcan.gc.ca/energy/efficiency/buildings/research/optimization/recommissioning/3795

As with energy audits, funding may be available in some jurisdictions to assist with covering the costs of the RCx process. The utility should be aware of potential funding opportunities and guidelines or rules if considering RCx for the WWTP or any other utility facility.

Status

The following questions are designed to help the plant staff better understand the utility's status with respect to RCx. Staff are also encouraged to develop and consider other questions that may be specific to their utility:

- Does the utility have a regularly occurring, comprehensive operations and maintenance plan that includes detailed verification of building systems function at the facility?
  - Verification of building systems function include balancing of flows, verifying schedule of equipment operations, and confirming that various types of motors (fans, pumps, or dampers, for example) and valves are functioning as expected.
- Is the facility equipped to automatically monitor the status and trends of the building systems (by a building automation system or other), and is the data produced regularly reviewed by personnel?
  - Having someone who is familiar with these trends and data will increase the chances of catching discrete energy-wasting problems when they first occur.

Action

If areas of the RCx assessment are considered Partially Optimized or Not Optimized, the following steps are recommended for developing and implementing an action plan.

- If both of the status questions are considered to be Optimized and Documented, the facility may reveal less opportunity for improvement during a RCx exercise. Having one completed, on occasion, may be a good method of augmenting personnel's knowledge. Potential action items may include
  - Updating the training of utility personnel to include appropriate RCx knowledge,
  - Review and modification of operations and maintenance procedures to include RCx facets that were previously excluded, and
  - Consider conducting occasional RCx exercises. Page 91 of the NRCan guide mentioned earlier contains a series of questions to help determine the need for completion of RCx exercises.
- If the status questions are considered to be Partially Optimized or Not Optimized, consideration for scheduling of an RCx exercise is recommended. Potential action items may include
  - Identifying helpful resources in the local area (energy saving organizations, consultants, or other),
  - Tailoring an RCx scope of work that is specific to the facility and needs,
  - Proceeding with the RCx work,
  - Implementing the recommendations from the RCx report, and
  - Consider conducting regular RCx exercises.

- To help prioritize the facility's buildings for completion of RCx exercises, a prescreening tool can be used.
    - A simple example is available from the Government of Canada (NRCan), titled the recommissioning (RCx) prescreening tool. This tool may be accessed using the following link: nrcan.gc.ca/energy/efficiency/buildings/research/optimization/recommissioning/11914
- A more involved benchmarking program is also available from Energy Star, called Portfolio Manager. Additional information about Portfolio Manager may be accessed using the following link: energystar.gov/buildings/facility-owners-and-managers/existing-buildings/use-portfolio-manager
    - There is also a WWTP-specific program within Portfolio Manager, which allows users to calculate an Energy Star rating specifically for wastewater treatment facilities. Additional information about this wastewater-specific tool may be accessed using the following link. The tool takes into account process-specific information and loadings to determine a score for the facility: energystar.gov/buildings/tools-and-resources/energy-star-score-wastewater-treatment-plants

Energy Management

Understanding

Energy management entails continuous engagement of the organization's people to manage energy usage. In its simplest form, it may consist of a utility green committee, leading the way to ensure that employees turn off lights and computers at the end of the day to realize the associated energy savings. In its most sophisticated form, there would be staff and resources dedicated to actively managing all energy usage.

A variety of industry standards exist that cover the area of energy management. International Organization for Standardization (ISO) 5001 and European EN 16001 both are energy management standards. It is thus possible to be ISO certified for energy management in addition to the more typical ISO certified quality standard. Energy management is also incorporated into several AWWA management standards. The Association of Energy Engineers has a certified energy manager professional designation. Bearers of the designation have passed an exam and have documented experience in the field. Additional information about this designation may be accessed by using the following link: aeecenter.org/i4a/pages/index.cfm?pageid=3351

While energy management certification is not a requirement for wastewater professionals, basic knowledge of energy standards can be helpful in beginning to implement plant-wide energy management initiatives and programs.

The USEPA provides a variety of resources for energy management at wastewater utilities, including a wastewater management fact sheet for energy conservation and the *Energy Management Guidebook for Water and Wastewater Utilities*. These resources include discussions of energy management techniques and examples of process improvements. USEPA resources are typically available online and may be downloaded at no charge.

Similar resources are available through NRCan. A compilation of NRCan's energy management best practices documents may be accessed using the following website link: nrcan.gc.ca/energy/efficiency/buildings/capacity-building-resources/learn-more/4259

Wastewater utilities are encouraged to include energy management activities in their daily operations. Even small actions, such as turning off lights, computers, and other electrical equipment, may amount to larger savings if they are consistently practiced by all employees on the site.

Status

The following questions are designed to help the plant staff better understand the utility's status with respect to energy management. Staff are also encouraged to develop and consider other questions that may be specific to their utility:

- Is there an active energy manager on site?
- Is a team responsible for leading energy management activities on site (i.e., a green team or similar group)?
- Does the utility participate in any local or citywide energy management initiatives?
- If these items do not exist, is there any staff member with the potential to be an energy manager?
- If these items do not exist, can a consultant potentially be engaged to provide this service?

Action

If areas of the energy management assessment are considered Partially Optimized or Not Optimized, the following steps are recommended for developing and implementing an action plan:

- Identify an individual who can be responsible for managing site energy management activities.
- Develop a group or team that can be responsible for leading site energy management activities.
- Collect data and information to help quantify current energy management practices at the site.

- Set energy management objectives and encourage collaboration to meet them.
    - Start small: Even a minor initiative to turn off lights and/or electronic equipment after use can result in measurable improvements if it is practiced consistently among utility staff.
- Continue to provide resources and tools to support the current energy manager or energy management team.
- Provide training to potential energy managers and/or teams.
- Consider hiring the services of an energy management consultant to develop an energy management plan, initiate an energy management team, or lead site energy management practices.

Energy Star has an energy management guidebook, titled *Guidelines for Energy Management*, to help energy managers and teams get started with the process. This resource may be accessed using the following link: energystar.gov/sites/default/files/buildings/tools/Guidelines%20for%20Energy%20Management%206_2013.pdf

Energy Measurement and Verification

Understanding

Principles of energy measurement and verification can be used by energy managers to monitor performance and by energy service companies to validate energy savings. The techniques applied to measuring energy consumption become critical when a service provider has guaranteed the savings from a particular upgrade, for example. Energy measurement and verification principles can be used to verify that efforts and investments are translating into the desired energy savings. Otherwise, it can be difficult or impossible to determine if changes have been effective by simply trending monthly utility bills. The information obtained by carefully measuring and verifying energy usage can be used to monitor progress in energy savings and, to the trained eye, can signal abnormalities or malfunctions of systems or components. This can be a useful tool for operations and maintenance, quality control, and energy-efficiency personnel.

An energy measurement and verification plan includes careful planning of measurement locations, frequency of data logging, and instrument accuracy combined with an understanding of the impact of several external factors (weather or occupancy, for example). Energy measurement can involve significant instrumentation costs, and it is important to be strategic about what to measure and how. A variety of industry resources exist that provide additional information about strategies for energy management and verification.

- The New York State Energy Research & Development Authority has published a substantial document titled Water & Wastewater Energy

Management Best Practices Handbook, 2010, which is available for download using the following link: nyserda.ny.gov/-/media/Files/EERP/Commercial/Sector/Municipalities/water-wastewater-energy-management.pdf. This document describes the basic steps involved in building an energy program as well as best practices in energy management for wastewater utilities.

- The Association of Energy Engineers has a certified measurement and verification professional designation. Bearers of the designation have passed an exam and have documented experience in the field. Additional information about this professional certification may be accessed using the following link: aeecenter.org/i4a/pages/index.cfm?pageid=3356. The course is based on the international performance measurement and verification protocol, and this document can be a valuable resource for anyone interested or practicing in the field (see efficiency valuation organization).

While WWTP staff may not be required to be certified energy measurement and verification professionals, measurement of strategic parameters may help to quantify the effectiveness of steps taken to manage site energy consumption. Positive results obtained can be beneficial in encouraging staff to continue energy management activities and consider implementing additional initiatives to further reduce site energy consumption.

Status

The following questions are designed to help the plant staff better understand the utility's status with respect to energy measurement and verification. Staff are also encouraged to develop and consider other questions that may be specific to their utility:

- Is the energy consumption of the facility measured and verified?
- Is there staff available to monitor this data, summarize the results, and communicate them to utility personnel?

Action

If areas of the energy measurement and verification assessment are considered Partially Optimized or Not Optimized, the following steps are recommended for developing and implementing an action plan:

- Modern instrumentation facilitates understanding where energy is consumed. Investing in instrumentation and the labor hours to monitor it can provide the team with valuable information. If energy consumption is not currently monitored, consider investing in basic instrumentation to monitor the most significant performance indicators that may help to quantify the effectiveness of energy management activities and encourage further team activities.

- This information can be used to monitor progress in energy savings and, to the trained eye, can signal abnormalities or malfunctions of systems or components.
- Consider trending energy measurement information so that long-term trends may be developed for further analysis.

Building and Energy Codes and Standards

Understanding

Building projects are regulated by various types of building codes. In many areas, this includes energy codes. In jurisdictions that have not adopted energy codes, there may be a guideline that serves as a reference for good energy practices. In the United States, most states have adopted one of the releases of ASHRAE 90.1: Energy Standard for Buildings, and some states have written their own energy codes. The most recent release of ASHRAE 90.1 was in 2013. ASHRAE publishes books and standards and conducts energy research that is used worldwide.

The National Electrical Manufacturers Association provides a pictorial graphic of commercial energy code adoption across the United States. A map of code adoption may be accessed using the following link: nema.org/Technical/FieldReps/Documents/EnergyCodes-Maps.pdf

In Canada, many provinces have adopted the National Energy Code for Buildings 2011 (NECB 2011). Many programs still refer to the predecessor of NECB 2011: the Model National Energy Code for Buildings 1997 (MNECB 1997).

This link, on the National Research Council Canada website, provides a table showing code adoption, including energy codes, for Canada: nrc-cnrc.gc.ca/eng/solutions/advisory/codes_centre/code_adoption.html

Many codes allow users to choose between various paths, including prescriptive, prescriptive with tradeoffs, and energy modeling. Typically, contractors involved with facility design projects will be familiar with local codes and standards to which they must adhere. However, it can be helpful for utility staff, particularly those involved in facility and energy upgrade projects, to have some basic familiarity with applicable codes for the plant and its local region.

Status

The following questions are designed to help the plant staff better understand the utility's status with respect to building and energy codes and standards. Staff are also encouraged to develop and consider other questions that may be specific to their utility:

- Does the plant's jurisdiction have an energy code or guideline?
- Does the facility include buildings that were completed prior to the adoption of the current set of codes in effect in the area?
- If deficiencies exist, has a plan been developed to bring site facilities into compliance with applicable codes and/or standards?

Action

If areas of the building and energy codes and standards assessment are considered Partially Optimized or Not Optimized, the following steps are recommended for developing and implementing an action plan.

- Prior to commencing a renovation project, investigate what codes, standards, and bylaws might apply to the plant's jurisdictions.
- Understand and document what deficiencies may exist between existing site buildings and the current applicable energy codes.
- Understand and document which of these items the utility may be obliged to install once major renovations begin (plumbing and life safety, for example).
- Consult with local expertise or study the current energy codes and guidelines in the plant's area.
  - It may be determined that current energy code documents relevant to the plant's area contain information that is useful to improving the plant's systems operations and outcomes.

Third-Party Certifications: Appliances, Equipment, and Whole Buildings

Understanding

Third-party certifications covered in this section are typically voluntary programs to recognize and validate the efforts of a manufacturer or owner to achieve green or efficient targets. In some cases, jurisdictions may have adopted these third-party certifications as a requirement. For instance, a city may decide that all their new buildings shall be built to LEED or Green Globes standards.

An overview of various programs follows.

Appliances and Equipment

Energy Star is a USEPA rating program (run by NRCan in Canada) that helps buyers to quickly recognize energy-efficient products with labels such as the one displayed in Figure 5-2.

(Source: USEPA)

Figure 5-2. Example of an Energy Star label found on many energy-efficient products in the US

Energy Star allows products meeting the category's high efficiency threshold to display the Energy Star label—typically the top 15 to 30 percent of the category. A wide range of consumer products are included in these rating programs, including heat pumps and water heaters. This link provides a list of categories: energystar.gov/products

Energy Star also has a category for homes, buildings, and WWTPs, which is discussed in various other sections of this chapter.

EnerGuide ratings on appliances in Canada are a valuable tool when purchasing energy-consuming appliances, such as refrigerators and clothes washers, for which the labels are mandatory.

It allows shoppers and purchasers to compare kWh of energy used per year between competitors and variations of the product as some appliance are divided by specific type or size groups.

Energy consumption in kWh may be converted to dollars per year cost by multiplying the kWh energy consumption by the applicable utility rate for electricity, for example:

$$554 \text{ kWh/year} \times \$0.10\text{/kWh} = \$55.40 \text{ per year energy cost}$$

An example EnerGuide label for a refrigerator (class 5A) is displayed in Figure 5-3.

The EnerGuide label displays four items: the annual energy consumption of the model in kilowatt hours (kWh); an energy consumption indicator, which positions the model compared with the most efficient and least efficient models in the same class; the type and capacity of models that make up this class; and the model number. The model displayed on this label consumes 554 kWh per year, compared to the best performing model, which consumes 481 kWh per year. When selecting items, such as appliances, it is also important to consider needs,

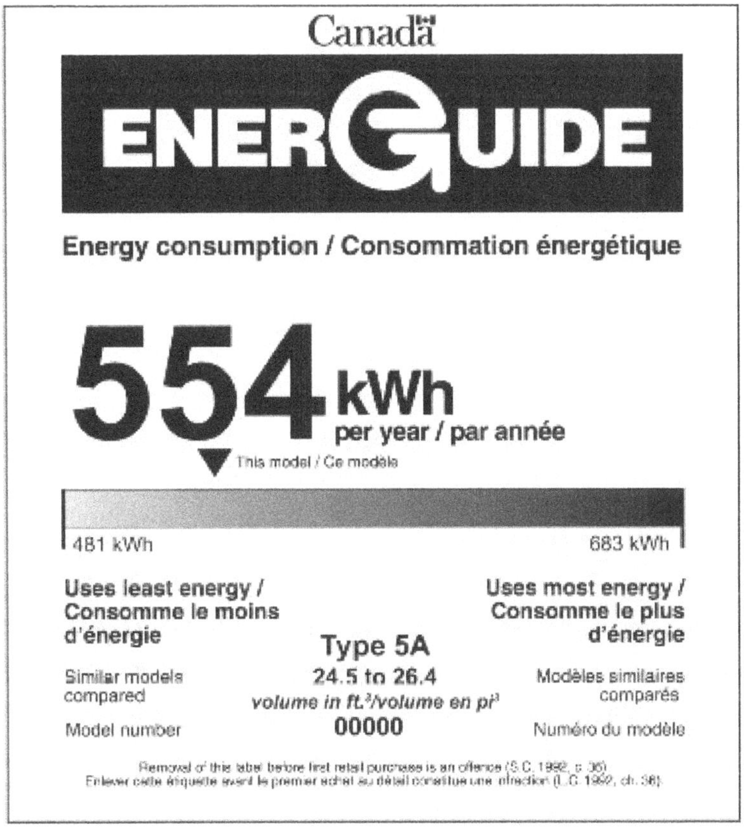

(Source: NRCan)

Figure 5-3. Example of an EnerGuide product label

such as the size of the refrigerator or other appliance being selected. The model displayed on the label has a capacity of approximately 25 ft^3. However, smaller models, with a capacity of 18 ft^3, can consume significantly less energy and be more cost effective to run.

HVAC Equipment

The Air Conditioning, Heating and Refrigeration Institute (AHRI) provides engineers and the public with independently tested efficiency data on a wide range of air conditioning, heating and refrigeration products. The directory can be found by accessing the following link: ahridirectory.org/ahridirectory/pages/home.aspx.

The institute also publishes standards and guidelines that are used throughout the world. Participation in the certification program is voluntary. However, most major manufacturers publish AHRI performance data on their technical data sheets.

Whole Buildings

Several options are available for certifying a green building to third-party standards.

LEED is a rating system that many now recognize. The program is used in 150 countries. Its use is mandated in certain jurisdictions for new construction. LEED also has a rating program for existing buildings. LEED certification provides independent, third-party verification that a building was designed and built using high performance strategies for human and environmental health. (For more information, see USGBC.org and in Canada, CaGBC.org.)

Green Globes is an online green building rating and certification tool used primarily in the United States and Canada. It can be applied to existing buildings as described in the following link: thegbi.org/green-globes-certification/how-to-certify/existing-buildings

The website also has technical and summary guides that can be downloaded free of charge. The certification process is a web-based questionnaire to be answered by the project manager and design team.

Energy Star has a building certification program that allows buildings that perform better than 75 percent of similar buildings to bear their mark. The Energy Star score varies from 1 to 100, and the building must achieve 75 or higher to be eligible. Performance is verified on an annual basis. Additional information about Energy Star recognition for buildings or plants is available by accessing the following link: energystar.gov/buildings/about-us/how-can-we-help-you/recognition/earn-recognition-your-building-or-plant

This Energy Star program makes use of Portfolio Manager. The program has many facets, one of which is specific to WWTPs in the United States. The Water Research Foundation provided benchmarking data to make the program relevant to WWTPs, and a great deal of effort went into the development of this tool. It allows comparison of WWTP EUI with expected plant EUI and then assigns an Energy Star score to the plant. The program is fairly easy to use, but it requires knowledge of several technical details and key data to be entered prior to obtaining a result. Additional information about the Energy Star program as it relates specifically to WWTPs may be accessed by using the following links:

- energystar.gov/sites/default/files/tools/Wastewater.pdf
- energystar.gov/buildings/tools-and-resources/energy-star-score-wastewater-treatment-plants

BOMABEST® is a national program launched in 2005 by The Building Owners and Managers Association Canada. It has an array of online assessment tools, and the four-level performance certification program is commonly used by commercial, government, and retail space. Additional information about the BOMABEST program may be accessed by using the following link: bomabest.com.

Status

The following questions are designed to help the plant staff better understand the utility's status with respect to third-party certifications. Staff are also encouraged to develop and consider other questions that may be specific to their utility:

- Appliances' efficiency is steadily improving. Are buildings on the WWTP site full of old appliances and equipment? If so, has a plan been developed to replace inefficient appliances and equipment over time?
- Has the utility adopted a goal that includes third-party certification at any level?
- Does the performance of the WWTP's buildings meet thresholds of third-party certification for energy performance (for instance, Energy Star, or whole-building systems, such as LEED or Green Globes)?
- Has the plant or utility adopted a policy such that new product purchases in eligible categories are rated by Energy Star, EnerGuide, or another appropriate entity?

Action

If areas of the third-party certification assessment are considered Partially Optimized or Not Optimized, the following steps are recommended for developing and implementing an action plan.

- Evaluate how the older portions of the facility may be improved in comparison to the jurisdiction's adoption of mandatory third-party certifications or the facility's identified internal target.
- Discuss how the utility can potentially make use of these third-party certifications. Adopt and incorporate the conclusions into the utility's vision, mandate, and procedures.
- Create procedures to guide facility managers toward energy-efficient purchasing of appliances, equipment, and renovation or new building design services.

Basics Metrics Sheet

An Excel spreadsheet is provided to Partnership for Clean Water subscribers, containing basic equations to enable those without energy modeling expertise to experiment with energy data. Users may input variables, such as pump hp and hours of operation, to estimate the potential impact of changes on energy consumption and power costs. The spreadsheet is provided in an unlocked format to give the user the flexibility of creating tables with the same equations. However, this freedom can result in functionality of the equations being broken; thus, it is recommended to save often and to save an original copy.

Each tab presents a component of energy-consuming systems, and instructions on use are provided within the spreadsheet.

While it is not recommended to make decisions based on such limited analysis, it can help identify areas of interest and areas that may have a high potential for cost savings as well as to help develop a basic understanding of how energy usage translates to dollars.

Conclusion: Facility Energy Optimization

This chapter provides an overview of the multiple facets of energy efficiency, and offers opportunity for further study and training. The chapter contains self-assessment questions for the self-assessment team to consider and suggested actions for developing and implementing an action plan related to facility energy optimization. While many of these changes will not affect effluent quality, they may have a significant impact on the facility's energy costs, depending on the nature of the changes that are implemented.

When implementing changes, it is important to clearly define the scope of the project, particularly when working with outside contractors. Establish priorities by using staff knowledge. Known information and data can help further shape this strategic plan. Additional information about prioritizing performance limiting factors and developing an action plan is included in chapter 8. Calculating the EUI for each building may be appropriate at this stage as it can help prioritize or narrow the scope of work. Assigning a budget to fund these activities is important for internal and external resources (labor and materials). Developing the appropriate budget may be incorporated as an item within the utility's action plan. Without support from the WWTP management and/or without a budget, it is unlikely that anything will get done. A smaller initial budget to create the scope of work may be necessary in some organizations to be able to assign staff to do this planning. Finally, the strategic plan needs to include an implementation schedule. All Partnership for Clean Water action plan items should include a target completion date. As described in chapter 8, the action plan is meant to be a live document that is updated as tasks are completed and to reflect actual progress. A culture of continuous improvement will help ensure that the efforts continue.

Summary–Performance Limiting Factors

Table 5.2 summarizes factors related to the facility energy assessment. Check whether these factors are Optimized and Documented, Partially Optimized, or Not Optimized. Factors identified as Partially Optimized or Not Optimized will be prioritized in chapter 8 on identification and prioritization of performance limiting factors, with which an action plan can be developed and implemented that allows for optimization of these parameters.

Table 5-2. Facility energy optimization assessment

Self-assessment category	Questions for gauging optimization status	Optimization Status		
		Optimized and Documented	Partially Optimized	Not Optimized
Building energy performance assessment	Does the wastewater treatment facility currently quantify energy consumption using energy monitoring, measurement, and verification programs and portfolio management tools that provide this information?			
	Has the total number of buildings on the WWTP site been quantified?			
	Have all of the different fuel sources used by the facility been quantified?			
	Is process energy consumption metered separately from the energy consumed by building systems?			
	Is the appropriate information (e.g., utility bills) currently compiled in a format that allows utility staff to complete this exercise?			
Energy audit	Has the utility completed an energy audit for either a single building or the entire facility?			
	Has the utility defined the needs and scope of work related to energy audits and efficiency upgrades?			
	Is the utility familiar with local codes and bylaws that affect implementation of energy-efficiency upgrades at the facility?			
	Are any major building renovations planned at the facility? If so, are there energy-efficient upgrades that would be beneficial to implement during these planned renovations?			

Continued

Self-assessment category	Questions for gauging optimization status	Optimization Status		
		Optimized and Documented	**Partially Optimized**	**Not Optimized**
Energy audit *(continued)*	Are utility staff aware of programs applicable to the building or project that would provide financial incentives to improve energy efficiency? If so, are utility staff aware of the rules and procedures required for eligibility to receive financial incentives?			
	Have subject matter experts been identified to assist the utility in moving project plans forward in this area?			
Re-/retro-commissioning	Does the utility have a regularly occurring, comprehensive operations and maintenance plan that includes detailed verification of building systems function at the facility?			
	Is the facility equipped to automatically monitor the status and trends of the building systems (by a building automation system or other), and is the data produced regularly reviewed by personnel?			
Energy management	Is there an active energy manager onsite?			
	Is a team responsible for leading energy management activities onsite (i.e., a green team or similar group)?			
	Does the utility participate in any local or citywide energy management initiatives?			
	If these items do not exist, is there any staff member with the potential to be an energy manager?			
	If these items do not exist, can a consultant potentially be engaged to provide this service?			
Energy measurement and verification	Is the energy consumption of the facility measured and verified?			
	Is there staff available to monitor this data, summarize the results, and communicate them to utility personnel?			

Self-assessment category	Questions for gauging optimization status	Optimization Status		
		Optimized and Documented	Partially Optimized	Not Optimized
Building and energy codes and standards	Does the plant's jurisdiction have an energy code or guideline?			
	Does the facility include buildings that were completed prior to the adoption of the current set of codes in effect in the area?			
	If deficiencies exist, has a plan been developed to bring site facilities into compliance with applicable codes and/or standards?			
Third-party certifications	Are buildings on the WWTP site full of old appliances and equipment? If so, has a plan been developed to replace inefficient appliances and equipment over time?			
	Has the utility adopted a goal that includes third-party certification at any level?			
	Does the performance of the WWTP's buildings meet the thresholds of third-party certification for energy performance?			
	Has the plant or utility adopted a policy such that new product purchases in eligible categories are rated by Energy Star, EnerGuide, or another appropriate entity?			

CHAPTER 6

APPLICATION OF OPERATIONAL CONCEPTS

Process control and the operator's ability to apply operational control concepts can be a significant factor affecting a utility's ability to optimize operations and performance. As a result, all plants completing the self-assessment process, regardless of the nature of the process control equipment used by the plant, should address the assessment questions in this chapter to identify performance limiting factors and develop an action plan for improvement.

In the first portions of the self-assessment, the level of plant performance is evaluated and a determination is made of design and process limitations that may be contributing to less-than-optimized performance. In this chapter, operational factors are assessed. Operational factors are those that relate to the unit process control functions. Significant performance limitations can often exist in this area. The approach and methods used in maintaining process control can significantly affect the performance of plants that have adequate facilities.

The heart of the operational factors assessment is sampling and associated process control testing followed by data interpretation and process adjustments. This chapter is divided into discussions of the following topics:

- Process control testing
- Operator application of concepts
- Communication
- Online instrumentation, process control system (PCS), and supervisory control and data acquisition (SCADA)

Inclusion of these topics in this chapter allows for assessment of features that are critical to maintaining optimized performance.

Process Control Testing

Understanding

For a wastewater treatment system to accomplish the goal of producing effluent that meets or exceeds permit requirements and can be returned to the environment or used for reuse applications, each unit treatment process needs to perform properly. This multiple barrier approach concept is discussed in chapter 2, and it is important because a breakdown in any one unit process places a greater burden on the remaining processes and increases the chance of harmful constituents, such as viable pathogenic organisms, reaching the receiving waters or having to be removed in subsequent reuse facilities. When each unit process performs properly, the benefits of providing multiple barriers are realized. To achieve these levels of performance for each unit process, information focusing on flows and loadings and on the performance of the various unit processes in the plant must be *routinely* gathered, recorded, and interpreted. Based on this information, the appropriate process control adjustments must then be implemented to maintain *consistent* process performance. The terms *routinely* and *consistent* are stressed because the plant should be adequately staffed, monitored, and controlled at all times when it is in operation to allow information to be gathered and for process control adjustments to be made whenever conditions require.

The first step in the development of a process control program is to set performance goals for each treatment process and to develop, at all levels within the utility, the tenacity to achieve them. The Partnership for Clean Water performance goals for treated effluent quality are provided in Table 1-1. Utilities are encouraged to adopt the PCW's performance goals although the selection of plant-specific goals for specific unit processes, plant effluent discharged, and additional critical parameters is also acceptable—and encouraged. For example, utility staff may choose to establish a goal of 2.0 mg/L dissolved oxygen levels in the secondary activated sludge basins and a secondary clarifier surface loading rate of 675 gpd/ft^2/. Similarly, utility staff may establish a goal of 12 days sludge age in the secondary process and an instantaneous maximum chlorine residual goal for the plant discharge monitoring location. Once the goals have been set, plant staff can use instrumentation to monitor performance relative to the goals and make the appropriate process adjustments.

The operating staff should then identify the controls available within the plant. A control is an adjustment that the plant staff has as an available option to implement a change that will affect the unit process performance. Examples of controls that typically exist in conventional activated sludge treatment plants are listed here by treatment process. These controls may not be applicable to all wastewater treatment plants.

- Influent flow
- Preliminary treatment

- Adjust or bypass screening process due to increased flows (rain event)
- Adjust flow through grit removal process to allow for improved grit settling
- Chemical addition
- Airflow if applicable

- Primary treatment
  - Number of clarifiers in service
  - Hydraulic and/or surface loading rates
  - Solids removal (increase skimmer rotation)
  - Chemical addition if applicable
- Secondary treatment
  - Number of basins and clarifiers in service
  - Hydraulic loading if applicable
  - Activated sludge mass in aeration basins
  - Activated sludge mass in secondary clarifiers
  - Chemical coagulant addition if applicable
  - Polymer addition if applicable
  - Air supply
  - Dissolved oxygen residuals
  - Return activated sludge
  - Solids wasting rates
  - Depth of sludge blankets
- Disinfection
  - Disinfectant type
  - Disinfectant concentration
  - Disinfectant application point
  - pH adjustment
  - Disinfection contact time

The development of specific controls that are appropriate for the unit processes that are applied at the utility is critical. An example of process control parameters for which a utility may set performance goals as a means to establish action levels for specific plant controls is displayed in Table 6-1. This is a general example only and may not reflect the desired control points or action levels that are suitable for every wastewater treatment plant configuration. Plants using tertiary or advanced treatment processes or striving to maximize resource recovery

Table 6-1. Example wastewater treatment plant process control parameters

Wastewater treatment plant unit process control parameters	
Sampling location/unit process	**Parameter**
Influent	Flow, MGD Total suspended solids, TSS, mg/L Biochemical oxygen demand, BOD_5, mg/L
Primary clarifiers	Flow, MGD TSS, mg/L BOD_5, mg/L BOD_5 mg/L removal efficiency TSS, mg/L removal efficiency Raw Sludge Removal, % Total Solids (%TS)
Secondary basins	Flow, MGD TSS, mg/L BOD_5, mg/L TKN, mg/L NH_3–N, mg/L NO_3^-–N, mg/L NO_2^-–N, mg/L Alkalinity, mg/L pH Temperature, °F O_2, mg/L O_2 utilization, #/hr, or #O_2/#BODr Loading rates, #BOD/KCF O_2 transfer DO, mg/L Number of basins in service Sludge age, days Mixed liquor suspended solids, mg/L RAS return flows, MGD RAS return concentration, mg/L
Secondary clarifiers	Loading rate, gal/ft^2/d Depth of blanket, ft WAS, MGD %TS RAS, MGD RAS/WAS
Disinfection/effluent quality	Flow, MGD TSS, mg/L BOD_5, mg/L NH_3–N, mg/L NO_3^-–N, mg/L NO_2^-–N, mg/L TKN, mg/L Total-P, mg/L Alkalinity pH Instantaneous Cl_2 residual Max Cl_2 residual Fecal Coliform, MPN (cfu/100 mL)

should set similar performance goals for these processes. Utilities should develop process control targets based on their specific performance goals, treatment processes, and specific regulatory requirements.

After controls are identified and goals are defined, a sampling and monitoring protocol should be established to provide information about the status of each unit process and if an adjustment needs to be made. For example, jar-test data can be used to determine the correct dosing for a centrifugal chemical feed pump, and data from a continuously recording flowmeter or dissolved oxygen sensor in the aeration basin can be used to determine when to adjust an influent gate or an air blower output. A quality assurance/quality control (QA/QC) program should be in place to ensure the accuracy of process control instrumentation and the validity of data, including the accuracy of communications with the plant's PCS. The operator should never be in the position of questioning whether the results of an online instrument are correct. The instruments need to be maintained and operated per the manufacturer's recommendations in order to provide reliable, accurate data at all times. Having no data is better than having wrong data.

Sample measurement techniques and locations are chosen on the basis of the information desired, the specific parameter being analyzed, and the frequency of sampling. For example, grab sample analysis may be used to analyze samples in which the measurements do not frequently change or are not needed on a frequent basis. Online analyzers are available for a wide variety of parameters and may be used to measure parameters in which the measurements change frequently or are needed often. Depending on the specific parameter analyzed, online analyzers have the ability to collect information and transmit it to the PCS on a frequent basis. Determining the frequency of sampling is based on several factors:

- The importance of the measurement in terms of its impact on process performance.
- How dynamic the value of the parameter is. The frequency with which the value changes will dictate the frequency of measurement.
    - Continuous monitoring is considered necessary and is highly recommended for critical parameters, such as flow, dissolved oxygen, and effluent disinfectant residual.
    - Continuous monitoring for selected parameters may be required by federal and local regulatory agencies.
    - Grab samples may be used for parameters that do not change as quickly and are deemed less critical to process control.
    - Composite sampling may be required to determine the weighted concentration of a contaminant.

Establishing robust standard operating procedures (SOPs) and QA/QC procedures is also critical for all sampling to ensure the quality of the data. For example, increased sampling frequency may be dictated when conditions are rapidly changing (e.g., collection system discharge events that impact influent water quality). This might be accomplished by using automatic samplers that initiate more frequent sampling at the beginning of an event. Establishing a protocol for the communication of process control data is critical to staff's ability to understand plant conditions at all times. Spreadsheets or other data collection tools could be established to serve this purpose. The implementation of any data communications method should also include training plant staff in its correct use.

Data collected continuously needs to be archived and reviewed frequently with plant staff. Consider the frequency of data collection to ensure that it is collected at a frequency that best meets the plant's needs for process control—data collected at too frequent an interval will not necessarily improve process control, and insufficient data will hinder process control efforts. Trend charts, displaying process control data in a graphical format, can be prepared to assist in the interpretation of the data. Many process control systems have the ability to quickly and easily graph various data points, improving the ease with which data can be viewed, interpreted, and correlated. The PCW's data-collection spreadsheets can also assist in trend chart development. Example trend charts include the depiction of dissolved oxygen concentrations over the length of a secondary basin so that any variability may be visualized and addressed. Level lines and consistent data, well within the established upper and lower control limits, indicate stable unit process performance. The data need to be interpreted and the necessary controls adjusted to continuously maintain each unit process at desired performance goals.

All of the administrative and operating personnel within the utility must understand the importance of the process control program in attaining the plant's performance goals and demonstrating a consistent commitment and tenacity to use and support the process control program. Development of performance goals by the utility's management without input from the frontline operations staff may not achieve the buy-in necessary for consistent and proper use of the process control program. If the frontline operators develop the program without management involvement, however, there may be administrative decisions made that are in conflict with the goals of the process control program. Many plants have instrumentation and control experts to assist in developing safety nets within the process control system to help protect the plant.

All of the operations staff should be involved in the development of the process control program. This helps to ensure that all relevant issues are addressed, that everyone understands the program, and that everyone will commit to use it. Management must give their support for the staff time required to ensure this involvement.

Finally, all components of the program must be documented in operating guidelines and SOPs. Example SOPs are included in the appendix. SOPs should include the appropriate stepwise procedures, including information about control and maintenance procedures and strategies. These guidelines should be written and tested by the operations staff. Once they are drafted, they should be readily accessible to the entire plant staff, and staff should receive training on appropriate and relevant SOPs. It is important to create an environment in which these guidelines are seen as living documents, meant to be updated and modified as new insights into the process control program are discovered.

Status

To assess the status of process control testing as it relates to producing consistently compliant effluent water, review the following items:

- Have performance goals been established for *each* treatment process in the plant?
  - Performance goals should be established for key unit treatment processes and used by the plant staff to focus plant operational activity. Any deviation of unit process performance from its established targets should result in rapid response to achieve and maintain the desired performance levels.
  - Procedures should be established that link process performance with actions associated with a given analytical result.
  - A clarifier sludge blanket depth target may be necessary for a plant that is consistently challenged to meet secondary clarifier TSS effluent performance goals. For example, when the secondary clarifier experiences rising solids from the sludge blanket, staff may choose to increase solids removal from the clarifier by either increasing or decreasing the RAS flow rate. Deciding whether to increase or decrease the RAS flow rate is best determined by completing a special study. This would decrease the potential for denitrification in the clarifier and solids in the clarifier effluent.
  - Staff should be aware of the interdependencies of this effort, as increased solids removal would also lower the RAS/WAS concentrations significantly. This type of target is site specific and depends on secondary clarifier effluent quality variations, plant capabilities, and solids handling facilities. SOPs should be developed for these conditions so that operators are familiar with how to react, particularly if these types of events do not occur frequently.

- Has a process control sampling and testing schedule been developed and implemented?
  - A sampling and testing schedule should be developed and applied that provides information essential to monitoring and adjusting plant controls.
  - The frequency of monitoring many of these parameters may be site specific and may be determined by specific regulatory requirements pertaining to the plant.
- Have monitoring spreadsheets or similar data collection tools been developed and implemented that allow data to be collected and interpreted?
  - Spreadsheets should allow collection and interpretation of data that can be used as a basis for process control decisions.
  - Data should be trended to allow for quick review and correlation with other daily parameters (e.g., total suspended solids, total solids, dissolved oxygen, biological oxygen demand, chemical dose, plant flow).
- Has an emergency response procedure been developed for loss of disinfection chemical feeds or for unacceptable plant effluent treated wastewater quality?
- Has a comprehensive activated sludge control strategy been developed, documented, and implemented?
- Have all of the plant staff been involved in the development of the process control program, and have they developed operational guidelines, including emergency response guidelines?

Action

If review of the items in the status section indicates that a complete process control program is not being conducted to support the plant's performance goals, action plans should be developed to address the identified limitations, and changes should be implemented to improve performance. The approach used by utility staff in developing the elements of a process control program is also important. As previously discussed, all appropriate utility staff should be involved in the development of the program, and the program should be documented in operational guidelines that are used and modified by the operations staff.

Depending on the current process control program in use, additional training related to the specific process control strategy chosen may be required. This training should focus on providing hands-on experience that provides all of the frontline operators with the confidence to make process control adjustments as appropriate when the raw wastewater quality changes. In developing the plant's process control program, consider the following:

- Setting performance goals for plant operation, developed in a manner to ensure a tenacious commitment by staff at all levels to achieve them
- Implementing control programming to support performance goals and protect treated effluent quality—programming should include proper alarm set points established in the PCS and creation of appropriate notifications for exceeding them.
- Creating a process control sampling and testing schedule to be used at the plant
- Monitoring spreadsheets to record and interpret process control data
- Establishing a dosage control strategy for other treatment chemicals
- Developing an emergency or failure plan, including a timeline for how the frontline staff will handle emergencies and plant process failures, including notification as required

Operator Application of Concepts

Understanding

Improvement of plant performance is ultimately achieved by implementing process control procedures that can be used to move a capable plant to achieve the desired final plant effluent goal (see Figure 1-1). Successful process control of a wastewater treatment plant involves producing a consistent effluent that meets or exceeds permit requirements and can be returned to the environment or used for reuse applications. To have this ability, an operating staff must be able to use appropriate unit process control tests to interpret water quality changes and use the results to make changes in unit process operation to meet performance goals. Refer to Table 6-1 for an example of typical unit process control parameters. To accurately feed chemicals, an operator must be able to complete the necessary calculations or have a spreadsheet that can easily convert the desired dosage to a pump feed rate in mL/min, L/s, or gpm and verify pump performance. If dry chemical feeders are used, the operating staff must also be able to calculate feed rates in lb/d or kg/d, calibrate the dry feeders, and make appropriate stock solutions. The pump can then be set at the required rate to feed the desired dosage. Chemical feed activities can be enhanced and more accurately controlled by flow pacing the chemical addition step. This approach uses flowmeters within the unit process to speed up or slow down the chemical pump to maintain a consistent dose.

Each operator must be proficient in these process control skills so that the process flow stream can be continually monitored and appropriate responses can be made on a timely basis. It is especially important that operating staff maintain their unit process control skills as complacency has been identified as a contributing factor to suboptimal process control and treatment plant performance.

Complacency can result in an inability to respond to rapidly changing conditions and can result in a loss of control of plant performance. It is important for operators to remain vigilant and aware of the potential possibility of changes in influent quality that may affect plant performance.

While process control is arguably the most important activity to achieve consistent, reliable plant performance, operators must also understand and apply other treatment concepts to ensure that performance goals are met. Beyond understanding and applying the technical concepts of wastewater treatment, the operating staff must also have the tenacity for consistently producing a compliant treated effluent using the existing facilities and must have the confidence to make the required operational changes to achieve this goal. The confidence to change an existing practice allows the staff to respond to variable conditions. For example, a staff with confidence will not hesitate to make a process change in response to changing plant influent water quality or to evaluate the use of a backup plan when conditions warrant. Commitment to achieving treated effluent that consistently meets or exceeds permit requirements and can be returned to the environment or used for reuse applications and the confidence to change are important characteristics of an operating staff possessing the tenacity to achieve their performance goals. Operators need to have knowledge of and experience in how to balance treatment changes with their important responsibility of protecting the environment and public health.

Status

The status of the operations staff's ability to understand and apply wastewater treatment concepts so that the plant produces effluent that meets or exceeds permit requirements and can be safely returned to the environment or used for reuse applications may be assessed by reviewing the following items:

- Do all operating staff possess the following characteristics?
  - The tenacity to achieve the plant's performance goals
    - When plant performance starts to deteriorate, is all plant staff proactive to investigate potential causes and take action to address the upset condition until performance improves?
  - A willingness to take responsibility for plant performance
    - When plant performance starts to deteriorate, does the front-line operations staff proactively assume responsibility to identify potential causes and correct the problem, or do they rely on supervisors to tell them what to do?
  - A willingness to learn
    - Are the plant staff interested in and willing to learn new ideas and process control procedures that support the plant's performance,

or is there an attitude that current practices are acceptable and nothing needs to change?

- ▲ Confidence to make changes in treatment
  - ▪ Do the plant staff have the knowledge, skills, and abilities to investigate alternative treatment strategies (e.g., a special study performed to identify techniques to improve solids settling)
- ▲ A tenacity to investigate process upset conditions and the confidence to implement change in an existing practice to optimize performance.
- ▲ Empowerment to make changes in treatment
  - ▪ Does the plant staff feel that they have the support of management necessary to make needed process control adjustments, or are their actions limited by a perception that they do not have the proper authority?
- ▲ Understanding of when to call for help and whom to call
  - ▪ When plant performance continues to deteriorate after the plant staff has made changes, are they willing to call for assistance, do they know where to call for that assistance, and do they know the point at which to make the call? Are triggers in place within the operational guidance to prompt operators to call for help?

- Can the plant staff conduct the necessary calculations or have easy access to computer programs or other tools to determine the appropriate chemical feed rates (e.g., mL/min, L/s, or gpm) necessary to deliver the required chemical dosage throughout the plant?
  - ▲ Assess if the operating staff can convert a desired chemical dose (e.g., mg/L or lb/mg) to a corresponding pumping rate (e.g., mL/min, L/s, or gpm) and vice versa.
- Can plant staff respond appropriately to changes in influent quality?
- Does the plant staff have the knowledge and skills to repair and/or maintain process components (pumps or valves, for example) and understand the benefits of and tasks associated with a preventative maintenance program?
  - ▲ Consider the implementation of an asset management program or maintenance management software to track asset condition, record maintenance activities, and more effectively plan for future repair and/or replacement activities.
- If proprietary processes are in use at the plant, have the plant staff worked with the manufacturer to determine the most appropriate controls to identify and address process upsets?

Action

If the operating staff does not understand or apply the wastewater treatment concepts as determined by a review of the items in the status section, action plans should be developed to address the identified limitations to ensure that improvements are implemented.

- If the plant staff does not have the tenacity for supplying effluent that consistently meets or exceeds permit requirements and can be safely returned to the environment or used for reuse applications on a continuous basis, consider ways to create an environment conducive to developing a staff with the characteristics as previously described. Ensure that the plan to achieve this objective incorporates management acceptance of the performance goals and a commitment to empowering frontline operators to make process control adjustments.

Communication

Understanding

Effective communication is the cornerstone of any successful organization. A utility should place high priority on dissemination of information to ensure it is properly conveyed and comprehended throughout all levels of operations to maintain optimized treatment performance. Communication flows in a multitude of directions; including horizontally, from peer to peer, vertically from management down, and from the front line up within the organizational structure.

Communication is usually classified as either formal or informal. Formal communications are most commonly recognized as informational or instructional and take place within a framework of the utility's established administrative structure. This type of information transfer may be accomplished through verbal means, such as through meetings with a group or individual, or written as in policies, SOPs, posted notices, emails, or electronic media. Informal communication may contain elements of formal communication but is generally recognized as bringing about the greatest teamwork and is the most common form used in day-to-day operations. The effectiveness of informal communication is essential as many workforces depend on this type of information transfer between employees to accomplish plant tasks, such as working safely, maintaining suitable treated water quality, evaluating potential treatment issues, and reviewing equipment operational status and maintenance requirements.

An often-overlooked component of communications is a check for comprehension. Most organizations are aware that it is important to provide information and instruction to their employees as well as arrange a means for the employees to voice their needs and concerns, but many overlook that it is just as important to check for understanding. Misunderstandings and differing interpretations among individuals often lead to unintended consequences. Accurate comprehension will ensure that actions are taken in accordance with given directives, procedures, and established goals.

Just as there are procedures for plant operation and maintenance, having similar communication procedures can benefit treatment plant operations. For example, developing procedures that allow plant staff to better understand who to contact, under what circumstances or performance triggers, and in what format can help promote proper communications when they are needed, resulting in a rapid response to plant issues. The establishment of communication forums, such as tailgate talks or operations meetings can also help to promote expanded communications and information sharing in other areas.

Communication at a wastewater treatment plant has many potential challenges: different shifts, workdays, locational separation, and levels of understanding. However, with perseverance and tenacity, these obstacles can be overcome. Use all appropriate modes of communication, such as email, comprehensive log entries, face-to-face meetings, and possibly electronic blogs or websites to disseminate information that contributes to successful plant operation.

Status

When evaluating the effectiveness and status of the utilities' communications, consider and address the following questions:

- Has a formal communication protocol been developed?
  - ▲ Meetings should be scheduled at regular intervals with a planned agenda.
  - ▲ A common area for posting notices and information should be considered.
  - ▲ Disseminated information should be readily available to employees by various means such as email accounts or searchable intranet sites.
- Do operators effectively communicate with each other?
  - ▲ There should be a deliberate exchange between operators to ensure pertinent operational information is passed along from one shift to the next.
  - ▲ It is suggested that a designated time exists for operators to meet with each other and with the plant supervisor. These may consist of daily, weekly, or monthly meetings. These meetings should be scheduled in a manner that is compatible with the work shift schedule.
- Do operators and maintenance workers effectively communicate to ensure plant maintenance status is provided to the operators frequently and accurately?

  - ▲ One of the most vulnerable areas of plant performance is when atypical plant repairs or operations are taking place. It is critical that maintenance status is communicated to operators in a timely, efficient manner.
- Is information communication encouraged between plant staff?
  - ▲ The utility should encourage positive employee relationships to facilitate the exchange of information.
  - ▲ Operational status and work in progress should be communicated between departments and employees through written or verbal means. This may include log sheets, data collection spreadsheets or graphs, status boards, or briefing sessions with individuals or groups.
- Does the utility ensure that employees understand the information presented?
  - ▲ Administration and management should not assume that the employees have complete comprehension without seeking confirmation. Always allow for questions and, if appropriate, request the employee to restate the conveyed material.
  - ▲ Staff should be skilled in communication techniques to recognize verbal and nonverbal cues that confirm acceptance and comprehension.

Action

If it is determined that ineffective or limited communications exists within the utility, action plans should be developed to address this deficiency. Actions to consider implementing to improve communications include

- participating in training programs to enhance communication skills;
- evaluating meeting structure for effective leadership, content, and audience reception;
- facilitating opportunities or methods for information transfer between employees;
- developing a culture to encourage employee questions and inquiries and providing opportunities for employees to share issues and concerns;
- systematically reviewing written communications for validity and updating requirements;
- providing a structured system for information transfer (examples of this include, but are not limited to, log books, data spreadsheets or graphs, bulletin boards, memos, emails, tailgate sessions, and one-on-one meetings); and
- implementing a process to verify communication comprehension.

Online Instrumentation, PCS, and SCADA

Understanding

The use of PCS and SCADA is essential to address control and reliability from the collection system through the treatment process to the treated effluent. The PCS enables operators to view live and historical data, trend critical parameters, and create reports. The information provided by the SCADA and PCS systems is a critical tool that the staff uses to make appropriate operational decisions and document plant performance.

In this section, all field devices and any equipment controlled or parameters measured by the operational staff will be referred to as *instrumentation*. Operators use instrumentation to generate data that helps them manage the treatment process. Instrumentation and the associated data historian capabilities should be compliant with local regulatory requirements as well as the requirements of the plant staff. The decision to fully optimize, control, and manage PCS and SCADA systems efficiently will require input from all parties, including the operational staff.

Instrumentation and a control system are tools in the operator's toolbox that can be used to learn about a system and control operations. Information from the control system should not merely be collected and stored but should be evaluated and communicated to plant staff. This can be achieved using a variety of methods as described in the previous section, including tailgate meetings, shift change meetings, operational meetings, or electronically. Regularly examining and evaluating data generated by instrumentation allows plant staff to learn more about operations, develop changes that can improve operations, and develop control software settings that allow for improved process control.

Instrumentation provides information that can be used by the PCS to alert the operator of process conditions and serve as an early warning that treatment may need to change due to weather conditions or changes caused by a manmade event, helping operators to avoid complacency.

Although the complexity of control systems may vary from utility to utility, the benefits that SCADA and a PCS can bring to the treatment process are independent of system size. Regardless of utility size, the Partnership for Clean Water encourages the use of instrumentation, PCS, and SCADA to collect appropriate process information and apply controls that are critical to maintaining plant effluent quality. These are tools that can aid utilities in completing the self-assessment process and achieving optimization.

The operator of a treatment facility may be compared to a detective, proactively looking for the anomalies in the system rather than maintaining a reactive mentality to process changes. Well-maintained and reliable instrumentation helps the operator make timely decisions that lead to optimized plant performance on a consistent basis. Making the decision to add instrumentation may require planning for purchase or replacement, including redundant instrumentation and spare parts. As utilities develop action plans for improvement, consider that the addition of instrumentation and the equipment it controls may require long-term budgeting, a step that should be taken into account when developing action items associated with instrumentation. Budgeting consideration should take into account any costs associated with ongoing equipment maintenance.

Proper calibration, verification, and maintenance are required to ensure accuracy and reliable performance. It is important to follow the manufacturer's instructions for the type of instruments in use at the plant. If on-line instrumentation is not reliable and not producing accurate data, this can lead to poor decision-making, potentially putting the treatment plant and treated wastewater quality at risk. Operational staff must have confidence in their instrumentation so that they are able to make well-informed treatment decisions.

On-line instrumentation is often used in conjunction with grab sampling although one does not necessarily take the place of the other. There are often minimum frequencies at which an operator is required to perform laboratory analyses. For example, some utilities perform chlorine, dissolved oxygen, or pH analysis at 4-hour intervals to protect the plant from compliance concerns in the event of an instrument failure. These results should be recorded for evaluation and comparison with on-line instrumentation.

A well-maintained data historian should be able to capture readings from instruments, archive data, and compress the data to 15-minute intervals or intervals selected by the utility without gaps in the data collection. Maintenance records should also be kept in hard copy and/or electronically. Currently, data can be stored on a hosted site. Operators must remember that the utility is the responsible party for data collection, archiving, and ensuring the integrity of the data regardless of where it is stored. Data gaps may need to be accounted for by operator testing in the lab, which may be a regulatory requirement. If online instrumentation or its associated data collection system is down, the results of laboratory tests may need to be collected at certain intervals to ensure process control and regulatory compliance until the problem is resolved.

A regulatory audit may require the provision of documentation and records of calibration, verification, and maintenance to the regulating authority for online instruments in use. In the laboratory, logs are also to be maintained for calibration, verification, and maintenance of devices to verify they operate in an acceptable range as prescribed by the manufacturer. When appropriate, control charts should be developed and updated annually for each piece of analytical equipment.

In a smaller treatment facility, calibrations and verifications may be performed by a licensed operator who has a comprehensive understanding of instrumentation. In a larger facility, the calibration technician or specialist who works on instrumentation may not necessarily be a licensed operator. However, the person performing these functions should have an understanding of the importance of the instrumentation and its role in maintaining treated wastewater effluent quality. The individual who performs the calibrations and verifications is site specific, but the main objective of the process is instrumentation that is operating in an acceptable range and providing accurate data to PCS and the data historian.

The instruments and devices used by the utility may be located in a variety of locations, some of which may be a significant distance from the central computer(s). Communications may involve travel of the signal through an elaborate process from the field to the control room.

When completing the self-assessment, the ability for operators to understand the network in use is important. Operators should have an understanding of how the instruments send signals to SCADA and PCS and how control devices react to the signals although expert knowledge of the control system, programming, and advanced troubleshooting is not required to use the valuable information that a control system can provide.

As stated in previous sections, the operators must determine what data to collect and the most suitable frequency of data collection. Control system monitoring and data collection of parameters such as basin influent flow, units in service, and temperature can help operators determine optimal oxygen transfer in a secondary basin. The frequency of data collection may vary depending on regulatory requirements, typical rate of change, and the robustness of the hardware and software applications used for data collection.

Status

The status of instrumentation is extremely important for completing the self-assessment, communicating results to regulatory agencies, and for operational staff to maintain a proactive approach. Each treatment process addressed in this guide requires different instrumentation and data collection requirements for operators to maintain a confidence level with the treatment processes and the analyzers used to monitor them.

To assess the status of instrumentation as it relates to achieving a desired performance level, review the following status questions. However, the plant's assessment of online instrumentation should not be considered to be limited to these items. The following questions are applicable to the majority of utilities and treatment plants. Utilities should also consider developing additional assessment areas, covering the existing devices at a facility.

- Maintenance is required to ensure reliability of the instrumentation. Has a maintenance and calibration schedule been implemented for all instrumentation?
- Has the utility implemented a training program to train operators in the care and maintenance of on-line instrumentation? Has the utility implemented a training program to train operators in the use and operation of the PCS and SCADA system?

- Does the utility have clearly defined roles for who is permitted to modify and maintain the process control system, and are those individuals properly trained?
- Application and placement: Are the instruments selected suitable for the application and physical placement for which they are installed? Does the placement of the instruments allow for ease of access for maintenance? Are environmental conditions suitable for analyzer placement? Is the placement of the sample collection and delivery lines suitable to obtain a representative sample of the source?
- Calibration and verification: Are the instruments calibrated and verified regularly? Do operators follow the manufacturer's instructions for calibration and verification? Are calibration and verification performed at a frequency that, at minimum, meets regulatory requirements? Are calibration records maintained and accessible when needed? Has an SOP been developed for the calibration and maintenance of all critical instrumentation?
- Are control systems tested and checked regularly to ensure proper operation and functionality when needed? This may include testing functionality of alarms and control capabilities.
- Redundancy: Are redundant instruments and equipment installed where appropriate? Does the plant have a plan established for the procurement of all critical parts to move toward making the plant bulletproof?
- Does the utility have provision for a backup server or control system in the event of communication loss or failure?
- Ensure SOPs, logbooks, instrument manuals, and work performed, such as calibration or verification, is completed and archived. Is all pertinent information documented? Is record-keeping adequate (i.e., is the necessary information maintained for the amount of time required by regulatory parties or internal plant processes)?
- Does the control system provide operational safety nets, consistent with plant practices and process performance goals to protect the operator and treated wastewater quality?

Action

If the instrumentation and control system are not adequate to support achieving the plant's performance goals, action plans should be developed to address the performance limiting factors and improve performance. The action plan for instrumentation should be included among the actions identified for individual plant unit processes.

If adequate maintenance and calibration are not being performed on instrumentation, develop a formalized schedule for completing this work, including responsible parties. Ensure that all staff is trained in the appropriate calibration and maintenance tasks and consider developing SOPs for these items.

To ensure operators are trained and confident in troubleshooting and operation of online instrumentation, develop a training program and SOPs utilizing internal and external subject matter experts as appropriate to increase instrumentation knowledge.

If instrumentation performance and maintenance is being affected by poor instrument selection or placement, select a more appropriate instrument or installation site. The distance between the sample and the analyzer should be minimized to prevent lag time in analysis. Instrument selection should be done on the basis of the sample type and expected concentration range for the parameter of interest, taking into account any additional issues, such as the potential for waste generation and disposal. Instruments should be installed in locations that are accessible for calibration, verification, and maintenance activities. Good judgment calls should be made when upgrading instrumentation to select the type of analyzer that is most suitable for the plant's application and intended use.

To ensure data validity, establish a procedure for instrument and data verification. This may include verification through the use of comparative grab sampling or purchased or prepared standards. When the results of on-line analyzers are compared with grab samples, utilities will typically define an acceptable percentage deviation for agreement (for example, 10 percent). This acceptable range may be parameter, instrument, and sample specific. If results do not agree within the acceptable deviation range, further troubleshooting, calibration, or maintenance activities may be required.

If insufficient spare parts are available onsite for critical analyzers, consider developing a plan to budget for and procure the required materials. Consider purchasing redundant instrumentation and/or equipment for critical parameters if this is required to achieve the plant's performance goals.

In cases in which record-keeping is insufficient or inconsistent, establish procedures for recording, maintaining, and storing records related to instrumentation. Ensure that this information is available to all operational staff and that staff is familiar with data recording procedures and data access locations. If records are stored in a variety of separate locations, consider integrating the information to allow information to be accessed and correlated more rapidly. Regulatory requirements may exist for data storage and record-keeping, and it is important that plant staff be familiar with these requirements.

Ensure that the control system provides safety nets to appropriately manage treated effluent water quality. Input from operational staff will be an important consideration in developing these controls. Be sure that the operational levels used in plant control programming are consistent with the plant's standard operating procedures. Consider the use of multiple levels of controls or alarms, such as utilizing a high alarm in conjunction with a high-high alarm, each with different associated consequences and control actions.

Instrumentation and control systems are important tools that can be used to improve operations, plant performance, and effluent quality. Staff may take a wide variety of actions to improve the way instrumentation is used and applied and how the data produced is collected, stored, evaluated, and communicated. While the self-assessment focuses on the most critical aspects of instrument use and those most applicable to a broad segment of wastewater treatment plants, utilities are encouraged to explore areas outside of these to develop an action plan to improve instrumentation performance in the most relevant areas.

It should be noted that even the most robust control system is simply only an aid to the operator in producing treated effluent that consistently meets the plant's permit requirements. The information that a control system collects and the processes it controls aid the operator, but the operator is still responsible for verifying the control system information by frequently checking equipment function, laboratory analyses, and other safeguards—a process that can be accomplished through plant walk-arounds and visual inspection.

Performance Limiting Factors Summary

Table 6-2 summarizes the operations factors that may be limiting performance of the plant. Check whether these factors are Optimized and Documented, Partially Optimized, or Not Optimized. Factors identified as Partially Optimized or Not Optimized will be prioritized in chapter 8 on identification and prioritization of performance limiting factors. Consider developing an action plan for improving any factor that was rated in either of these categories.

References

American Water Works Association. 2001. *M2—Instrumentation and Control Manual*. 3rd ed. Denver, CO: AWWA.

ANSI/AWWA G510, Wastewater Treatment Plant Operation and Management Standard Methods for Examination of Water and Wastewater. 22nd ed.

Jenkins, D., M.G. Richard, and G.T. Daigger. 2004. *Manual on the Causes and Control of Activated Sludge Bulking, Foaming, and Other Solids Separation Problems, Third Edition*. Baton Rouge, Fla: Lewis Publishers.

Tchobanoglous, G., H. D. Stensel, R. Tsuchihashi, and F. Burton. Metcalf and Eddy. 2014. *Wastewater Engineering: Treatment and Resource Recovery*. 5th ed. New York, NY: McGraw Hill.

Table 6-2. Operational factors assessment

Self-assessment category	Questions for gauging system adequacy	Response (check one category)		
		Optimized and Documented	Partially Optimized	Not Optimized
Process control	Have performance goals been established for each treatment process in the plant?			
	Has a process control sampling and testing schedule been developed and implemented?			
	Have monitoring spreadsheets or similar data collection tools been developed and implemented that allow data to be collected and interpreted?			
	Has an emergency response procedure been developed for loss of disinfection chemical feeds or for unacceptable plant effluent treated wastewater quality?			
	Has a comprehensive activated sludge process control strategy been developed, documented, and implemented?			
	Has all of the plant staff been involved in the development of the process control program, and have they developed operational guidelines, including emergency response guidelines?			
Operator application of concepts	Do all operating staff have the following characteristics? • The tenacity to achieve the plant's performance goals • A willingness to take responsibility for plant performance • A willingness to learn • Confidence to make changes in treatment • A tenacity to investigate process upset conditions and the confidence to implement change in an existing practice to optimize performance • Empowerment to make changes in treatment • An understanding of when to call for help and whom to call			

Continued

Self-assessment category	Questions for gauging system adequacy	Response (check one category)		
		Optimized and Documented	Partially Optimized	Not Optimized
Operator application of concepts *(continued)*	Can the plant staff conduct the necessary calculations or have easy access to computer programs or other tools to determine the appropriate chemical feed rates (e.g., mL/min, L/s, or gpm) necessary to deliver the required chemical dosage throughout the plant?			
	Can plant staff respond appropriately to changes in influent quality?			
	Does the plant staff have the knowledge and skills to repair and/or maintain process components (pumps or valves, for example) and understand the benefits of and tasks associated with a preventative maintenance program?			
	If proprietary processes are in use at the plant, has the plant staff worked with the manufacturer to determine the most appropriate controls to identify and address process upsets?			
Communication	Has a formal communication protocol been developed?			
	Do operators effectively communicate with each other?			
	Do operators and maintenance workers effectively communicate to ensure plant maintenance status is provided to the operators frequently and accurately?			
	Is information communication encouraged between plant staff?			
	Does the utility ensure that employees understand the information presented?			
On-line instrumentation, PCS, and SCADA	Has a maintenance and calibration schedule been implemented for all instrumentation?			

Self-assessment category	Questions for gauging system adequacy	Response (check one category)		
		Optimized and Documented	Partially Optimized	Not Optimized
On-line instrumentation, PCS, and SCADA *(continued)*	Has the utility implemented a training program to train operators in the care and maintenance of on-line instrumentation? Has the utility implemented a training program to train operators in the use and operation of the PCS and SCADA systems?			
	Does the utility have clearly defined roles for who is permitted to modify and maintain the process control system, and are those individuals properly trained?			
	Application and placement: Are the instruments selected suitable for the application and physical placement for which they are installed? Does the placement of the instruments allow for ease of access for maintenance? Are environmental conditions suitable for analyzer placement? Is the placement of the sample collection and delivery lines suitable to obtain a representative sample?			
	Calibration and verification: Are the instruments calibrated and verified regularly? Do operators follow the manufacturer's instructions for calibration and verification? Are calibration and verification performed at a frequency that, at minimum, meets regulatory requirements? Are calibration records maintained and accessible when needed? Has an SOP been developed for the calibration and maintenance of all critical instrumentation?			
	Are control systems tested and checked regularly to ensure proper operation and functionality when needed? This may include testing functionality of alarms and control capabilities.			

Continued

Self-assessment category	Questions for gauging system adequacy	Response (check one category)		
		Optimized and Documented	Partially Optimized	Not Optimized
On-line instrumentation, PCS, and SCADA *(continued)*	Redundancy: Are redundant instruments and equipment installed where appropriate? Does the plant have a plan established for the procurement of all critical parts to move toward making the plant bulletproof?			
	Does the utility have provision for a backup server or control system in the event of communication loss or failure?			
	Is all pertinent information documented? Is record-keeping adequate (i.e., is the necessary information maintained for the amount of time required by regulatory parties or internal plant processes)?			
	Does the control system provide operational safety nets consistent with plant practices and process performance goals to protect the operator and treated wastewater quality?			

CHAPTER 7

ADMINISTRATION

Administrative practices can significantly affect a utility's ability to optimize operations and performance. As a result, all plants completing the self-assessment process, regardless of configuration, should address the assessment questions in this chapter to identify performance limiting factors and develop an action plan for improvement.

The focus of this chapter is the assessment of administrative factors relative to plant performance. The administrative factors included in the self-assessment are listed below. This chapter is divided into sections that address each of these factors:

- Administrative policies
- Acceptance of optimization goals
- Involvement of all parties in the Partnership process
- Documentation/demonstration of addressing complacency
- Training
- Staffing
- Funding

Evaluation of administrative performance limiting factors is subjective and is done on the basis of management and staff involvement as part of the utility's self-assessment team. As described in previous chapters, the self-assessment team should optimally incorporate personnel from all disciplines and levels of the organization, including management. With management involvement throughout the self-assessment process, the questions and topics presented in this chapter should be addressed in an open fashion and serve to promote communication throughout the organization. This portion of the assessment should be performed following the capacity assessment and performance assessment. Budgeting and financial planning, staffing levels, and administrative policies are the mechanisms that plant owners and/or administrators generally use to implement utility objectives.

Administrative Policies

Understanding

Often, the impact of administrative limitations can be difficult to discern. The impact of administrative practices on performance or operations is not as direct as it is with the design or operational factors assessed in previous chapters of this guide. Additionally, it can be difficult to point fingers at the people who pay the bills, including salaries. Personalities and egos can also potentially interfere with assessment of administrative factors. Awareness of these significant influences must be maintained when assessing administrative factors. In order for a utility to strive for optimized performance and the highest levels of efficiency, a demonstrated commitment to excellence needs to be in place at all levels of the organization, including upper management. This commitment must be based on an understanding of the importance of optimization to effluent quality and plant efficiency as well as the impact of effluent quality on environmental protection. Administrators must be willing to pursue actions aimed at improving plant performance that consistently demonstrate an understanding of both the environmental implications associated with operating a wastewater treatment plant and of their responsibility in producing an effluent stream that consistently meets regulatory compliance requirements. The same actions that can help to optimize treated effluent wastewater quality to meet current regulations can also have the potential to improve efficiency and reduce operating costs. Administrators must be aware that excursions from optimized performance may result in noncompliance and associated fines as well as compromised quality of the receiving water.

Administrators must also understand that to optimize wastewater treatment plant performance, all unit processes must be performing at high levels on a continuous basis. The unit processes used at the plant provide multiple barriers that protect each downstream process as well as effluent quality. When upstream processes are optimized, this can help to reduce the loading on downstream processes and optimize the performance of each unit process at the plant. Accordingly, administrators should provide direction on developing goals for plant unit processes and effluent quality and emphasize to the operating staff the importance of achieving these goals. These goals should be understood and accepted by staff at all levels of the organization. Administrators should encourage establishment of effluent quality and energy-efficiency goals that provide a margin of safety with respect to regulatory requirements and empower their operating staff to achieve them. Utilities should consider adopting the Partnership for Clean Water optimization goals for treated wastewater effluent quality and energy efficiency (Table 1-1) as well as adopt relevant internal performance goals for additional plant unit processes not specifically addressed in this guidance manual. Both short- and long-term goals may also be established to work toward resource recovery as the highest level of plant performance whether it is something that may be achieved at the current time or in the future. Optimization goals should attempt to balance achieving a sufficient level of effluent quality while continuing to maintain operating costs at a reasonable level.

Typically, administrators verbally support goals of low cost, safe working conditions, good plant performance, and high employee morale. An important question that must be answered is, "Where does treated effluent wastewater quality fit in?" Often, administrators can be challenged by balancing the sometimes conflicting requirements for treated effluent wastewater quality versus minimizing costs, and this question can be answered by observing the items implemented or supported by the administrators. If major capital projects are implemented throughout the utility while the plant remains unattended and neglected, priorities regarding treated effluent wastewater quality and compliance may be being overlooked. Fortunately, improving treated effluent wastewater quality and reducing operating costs are not necessarily mutually exclusive as many of the self-assessment areas of this guidance have demonstrated.

Administrators should function with an awareness that they want to achieve a consistently compliant treated effluent wastewater as the end product of their treatment efforts and produce this effluent in as efficient a manner as possible. Administrators should strive to improve working conditions, support plant special studies and operation, provide adequate numbers of qualified staff, lower treatment costs, and meet other similar goals within the context of first achieving and maintaining compliant treated effluent wastewater. Lowering treatment costs should not be done at the expense of treated wastewater effluent quality, and many plants will discover that the optimization practices that result in improved effluent quality can also result in reductions in plant operating costs.

At the other end of the spectrum is an administrative attitude that borders on complacency: "We just raised rates last year, and we aren't willing to pursue additional revenues; besides no one is drinking the wastewater, and it's close to meeting the permit, so that's good enough." Administrators who fall into this category usually are identified as contributing to inadequate performance during an administrative assessment.

Utility administrators also need to be aware of the impact that administrative policies have on treatment plant performance. Consider an example in which a city manager forbids plant operators to waste sludge as frequently as necessary because operating the WAS pump causes increased biosolids disposal costs. This administrative policy has the potential to negatively affect treatment plant performance.

Minor modifications identified in an action plan by the utility staff during the self-assessment and the potential costs associated with these problems can often serve as a basis for assessing administrative factors limiting performance. For example, the plant staff may have correctly identified needed minor modifications for the facility and presented these needs to the utility manager but had their requests declined. However, the self-assessment team must solicit the other side of the story from the administrators to see if the administration is indeed unsupportive in correcting the problem. There have been instances in which operators or plant superintendents have convinced administrators to spend money to "correct" problems, only to find that no improvement in plant performance resulted.

Another area in which administrators can significantly, though indirectly, affect plant performance is through personnel motivation. A positive influence exists if administrators encourage personal and professional growth through support of training, using both in-house and external opportunities; involvement in professional organizations, such as AWWA and the Water Environment Federation (WEF); and tangible awards for initial or upgrading of certification levels. If, however, administrators eliminate or skimp on essential operator training; de-emphasize operator, laboratory, or related positions through substandard salaries; or otherwise provide a negative influence on staff morale, they can have a significant detrimental effect on plant performance.

Utility management needs to discourage complacency, especially when the utility may treat a consistent influent that varies little in its composition, quality, and quantity. Development of an environment that fosters a commitment to excellence can be the best defense against complacency. This requires the involvement of all staff members to create an empowered staff that can effectively respond to changing conditions and development of an appropriate training program to ensure that operator skill is maintained when response to any changes in treatment plant conditions is required.

Self-assessing how the utility's administrative policies actually support the plant's performance goals is potentially very difficult. One sign that changes in administrative support may be needed is if this portion of the self-assessment is prepared without input from the frontline operations staff. Another indication may be if a decision is made that no administrative problems exist and only limited effort is needed in this regard.

Status

The status of administrative policies as they relate to producing a consistently compliant treated effluent may be assessed by reviewing the following items:

- Do the utility's overall goals and commitment, at the highest level of management, indicate a focus on treated wastewater effluent quality and energy efficiency and include the role of the wastewater treatment plant and its operation in achieving these goals?
  - ▲ The utility should have a documented and followed mission statement identifying effluent water quality (i.e., regulatory compliance) as a utility priority with consideration of improved efficiency and/or resource recovery as a current or future utility goal.

- Does the utility have a strategic planning process? Such a process should include
  - financial planning,
  - long-range facility planning,
  - capital replacement plans, and
  - utility master plans.
- Does the operating staff have the authority to make required operation, maintenance, or administrative decisions?
- Do management styles, organizational capabilities, or communication practices at any management level adversely affect performance?
  - Management objectives should encourage empowerment of staff to make decisions.
  - Organizational structure should support achieving effluent quality and efficiency goals.
  - There should be clear alignment between plant goals and plant operation.
  - Utility staff should have regular meetings to review data and discuss operating issues, effluent water quality concerns, or other issues. This could occur at operator meetings that are regularly scheduled. This discussion may be more sustainable when a formal meeting is scheduled and a commitment is made from all stakeholders to attend and participate.
- Do administrators have a firsthand knowledge of plant needs through visits or discussions with utility staff?
  - Administrators should be aware of plant operational and effluent water quality issues.
  - Administrators should be accessible to utility staff—perhaps attending the meetings described above.
- Does the utility have ongoing public information activities, including customer outreach?
- Has the utility maintained awareness of existing and impending regulations, and has it used this information to prepare long-term plans to ensure compliance?
- Does management support utility staff involvement in professional organizations?
  - Staff should be encouraged to participate within the industry to learn new developments and be knowledgeable in current treatment advancements.

- Does management prioritize compliance over treatment costs?
  - The plant's effluent quality and compliance objectives should be achieved first, and then quantity and cost-effectiveness issues should be prioritized. In a noncompliance situation, the cost of fines has the potential to eclipse the costs associated with treatment and/or operations.

Action

If, after a review of the criteria in the status section, the utility determines that its policies do not support producing consistently compliant and good quality treated effluent wastewater, action plans should be developed to address the identified limitations, and improvements should be implemented to change attitudes throughout the utility. Some suggested means of achieving this are highlighted below.

- Develop a mission statement.
- Develop a charter. This charter can range from one that includes only operational staff at the plant to one that includes all employees in the organization. The key is to collaborate on which ideals need to be included in the charter and have all team members indicate their acceptance by signing the document. This then becomes a personal commitment to excellence.
- If, at any level in the utility, there is a lack of understanding of the need to comply with applicable wastewater treatment regulations or a lack of commitment to do so, develop an action plan to change these attitudes throughout the organization. Part of this action plan should address how this change of attitude will be communicated throughout the utility.
- The utility should make sure that there is a formal policy or process to assure that frontline operations staff have the authority and capability to change plant operations to systematically respond to periods of process upset or significant changes in influent flow or quality. This policy should also include procedures to follow when the ability to meet regulatory permit requirements may be compromised.
- The utility should support the above policy with a training program that ensures the empowered staff have the tools, compensation, knowledge, skills, and staffing levels necessary to respond to unanticipated treatment process changes or variability in influent wastewater quality.
- If there is a lack of understanding of existing and impending regulations, develop a list of requirements of these regulations and assess their impact on the existing plant.
- If the utility's management and operations staff perspectives significantly differ about the needs of the plant, they should develop an action plan

that clarifies and addresses current needs and develop procedures that facilitate staff input into the budgeting process.

- If administrative policies are found that inappropriately dictate operations decisions, they should be changed so that the plant's operations staff are empowered to make these changes at the appropriate level and after appropriate training.
- If communications procedures within the utility at any level are found to negatively affect process operation or treated wastewater effluent quality, improved formalized communication procedures, such as regular meetings or introduction of group decision processes, should be implemented.

Acceptance of Optimization Goals

Understanding

Wastewater treatment plant staff are encouraged to set goals that enable them to safely comply with all applicable regulatory requirements. The Partnership for Clean Water goals are listed in chapter 1, Table 1-1. The utility may also choose to initially develop internal goals as a starting point and later implement the PCW goals as plant operations and personnel continue to progress toward true plant optimization. Plants that are already performing at a high level may select to implement more aggressive goals than those of the PCW and may incorporate resource recovery and more aggressive energy-efficiency targets into utility goals. Regardless of how the goals are chosen, these goals should be reflected within operational guidelines and unit process limits. There should be clear alignment between the stated goals and the way in which the treatment plant is operated. Furthermore, this commitment to the defined goals needs to be evident even when treatment is challenging and process performance is tenuous. If goals are abandoned during times of difficult treatment, the utility needs to step back and evaluate if goals are understood and embraced by all operations staff. If there is clear understanding of the importance of the optimization goals as they relate to regulatory compliance, environmental water quality, and operational efficiency, there should be no resistance to moving the plant culture in that direction.

Status

To determine the status of universal acceptance of the optimization goals are within a utility, the treatment plant self-assessment team should evaluate and review the following questions:

- Does operational guidance reflect the optimization goals adopted by the treatment plant?

- Are plant SOPs aligned to support the multibarrier approach and plant optimization goals? There should be clear alignment between stated goals and plant operation and controls. If there is not, the importance of the goal can be inadvertently undermined or minimized.
- Are stated optimization goals enforced and followed even during periods when treatment is challenging or unanticipated events are encountered?
- Is there provision within the plant unit process operational guidance requiring operator intervention and/or action taken prior to unit process performance exceeding the optimization goals? It is important to identify, prior to reaching plant performance goals, actions operators shall take to evaluate the why and how to remain in control of the unit process.

- Do process control narratives of unit treatment processes incorporate the operational goals into the control system in the creation of safety nets? If the treatment plant has a control or SCADA system, an opportunity exists to develop safety nets for the operators that allow them to protect the plant. These should be implemented as appropriate.
    - Are alarms set up within the control system to give the operators advanced warning of flow or process loading changes?
    - Are there alarms set up within the control system to give the operators advanced warning of potential compliance excursions in effluent quality? For example, if the plant's effluent limit for chlorine is nondetectable, is a chlorine analyzer set to alarm if a detectable concentration of chlorine is present in the plant effluent? Operators may consider implementing different alarm warning levels, such as high alarms and high-high alarms, which would require varying levels of staff response.
- Does the plant operations staff maintain goals even during challenging treatment events that require process control adjustments to meet goals?
    - The true test of goal acceptance is during times when the unit processes are being pushed to meet regulatory compliance requirements. Does the plant staff and, more importantly, plant management, support operational decisions that value effluent water quality? If there is a conflict between the stated goals and what operators are expected to do when operations are challenging, there will be confusion as to what the priorities are for the operators. There should never be a question in the operator's mind about the top priority of plant operations.

Action

After careful evaluation of these questions, determine the best approach to ensure that the treatment plant optimization goals are clearly in alignment with plant policy and procedures, especially during times of challenging treatment. If there is a disparity between the goals and how the plant is operated, develop an action plan to meet with the treatment operations team and come to a resolution. This may require some difficult conversations with management, SCADA system modifications, or rewriting of operational guidance to achieve. The effort will be well worth it as the operational consistency and operator empowerment will far outweigh the effort required. Some actions to consider are to

- review SOPs and operational guidance, deliberately incorporating the plant goals into these documents;
- discuss with management the importance of developing and maintaining plant goals, emphasizing the importance of doing this even when challenges arise at the plant and gaining consensus on placing a priority on treated effluent wastewater quality and regulatory compliance; and
- be creative and develop safety nets within the control system to protect the plant and the operator. Control systems are some of the most underutilized resources available at the treatment plant to help accomplish this. Ensure these safety nets are set at levels that are in alignment with unit process goals.

Involvement of All Parties in the Partnership Process

Understanding

As described in previous chapters, the self-assessment team optimally should incorporate personnel from all disciplines and levels of the organization, including management. Involvement of all parties in the PCW process develops and nurtures personal and organizational ownership of and commitment to common goals that promote the best opportunity for success. Strong and committed leadership is an important key to establishing, cultivating, and directing this team effort to assure all parties remain committed to and focused on optimization goals. Utilizing the collective knowledge, experience, and wisdom of all parties is simply the best and most efficient way to achieve excellence. The ability to identify and successfully address and correct issues and provide guidance that enables all personnel to work to achieve the same objective will facilitate this process. The best opportunity for success requires administrators that involve staff at all levels to develop and set goals for the organization as well as promote an environment in which all parties are expected to participate in the optimization process.

Status

The status of the involvement of all parties in the PCW process may be assessed by reviewing the following items:

- Has the organization developed a vision and/or mission statement that incorporates the Partnership for Clean Water optimization philosophy?
  - The vision and/or mission statement should clearly highlight the core vision and mission of the organization. The development and periodic review of the vision/mission statement will ideally provide an opportunity for all employees to reflect on what is most important, offer input to define or refine the vision and/or mission, and begin or reinforce the process of involvement of all parties in the Partnership for Clean Water process.
- Is there strong leadership throughout the organization that supports the Partnership for Clean Water culture of excellence?
  - Leadership is a process whereby an individual (or individuals) influence others to achieve a common goal. Ideally, leadership exists at all levels of an organization. The ability to establish and cultivate strong leadership in the organization will help to assure that near- and long-term optimization goals are developed and met.
  - Effective leadership communication helps employees understand the utility's overall business strategy and how they contribute to achieving optimization goals by sharing information with employees on how the company as a whole and the employee's own division are working toward optimization.
- Have all personnel been briefed on the purpose of the Partnership for Clean Water program and do they have a good understanding of the overall intent to optimize treatment as well as the benefits of being a member of the Partnership for Clean Water?
  - Employees should be aware of the data collection required, its interpretation, and its use as a tool to measure progress toward optimization. One of the keys to obtaining buy-in is to quantify unit process performance improvements through the data interpretation process. The PCW provides software tools to subscriber utilities that help quantify process performance.
  - Employees should be encouraged and willing to seek out and identify performance limiting factors, recommend specific changes, and become actively involved in the development and implementation of action plans that mitigate or eliminate them. Remember the level of involvement of the team is directly related to them feeling they are a valuable part of the process and their voice is heard.

  - ▲ Employees should be trained to be vigilant and remain tenacious in their awareness of actions and situations that may adversely affect treatment goals and compromise optimization. These scenarios should be reviewed and proper responses documented.
- Do the employees feel that their roles are well defined and do they understand their roles in the optimization process?
- Do administrators set goals and provide an environment that expects and encourages involvement in optimization at all levels?

Action

If all parties are not involved in the partnership process as determined by a review of the items discussed in the status section, action plans should be developed to address the identified limitations. Development of an action plan may involve consideration of the following topics:

- Evaluate the specific ways the organization promotes, supports, and sustains a culture of excellence to ensure involvement of all parties in the Partnership process and revise if necessary.
- Develop or revise vision and mission statements involving all parties in the process.
- If there are problems with leadership and administration, identify the problem(s) and develop action plans to address them.
- If there is a lack of understanding of the partnership process and goals, develop action plans to ensure all parties obtain this knowledge through proper training and practice.
- Be sure to include members from all levels of the organization on the self-assessment team.
- Develop methods and training to cultivate institutional tenacity and awareness of complacency issues.

Documentation/Demonstration of Addressing Complacency

Understanding

Complacency may lead to nonoptimized and inefficient treatment plant operation. Complacency is defined as a situation in which inadequate capability exists to maintain optimized treatment performance during nonroutine events. The capability to address complacency involves skills, policies, and procedures intended to enhance problem solving and address rare and potentially serious events (RPSEs) that may lead to extreme and prolonged degradation in operations and treated effluent wastewater quality. The primary objective of documenting and addressing complacency is to demonstrate the utility's capability to address RPSEs, enhanced problem solving by frontline personnel so that taking action is not limited to managers or other indirect operations personnel that do not have proximity to quickly move to change processes and procedures when conditions call for change, and an internal ongoing process whereby developing situations are used to plan for required operational changes and past experiences lead to beneficial gains in anticomplacency measures.

Table 7-1 shows actions optimized utilities undertake to avoid complacency. A self-assessment examines all aspects of the plant operation, including problem solving and decision-making. Complacency may relate to the ways that wastewater treatment plant staff are equipped to handle significant changes in influent raw wastewater flow or composition. In addition, complacency may involve situations such as a treatment process upset and the development of sustainable procedures that can be implemented under these conditions to meet regulatory requirements in as practical and efficient a manner as possible.

Status

To assess the status regarding complacency, review the factors listed in Table 7-1, ask if any of these factors is not present at the utility, and address the status questions following the table. In addition, utilities should strive to identify RPSEs specific to the unique aspects of the wastewater treatment plant's service area, influent quality, treatment technologies, and/or regulatory permit requirements and to develop policies and procedures to mitigate or address events that may lead to nonoptimized treatment performance.

- During the past 10 years, how has plant staff responded to a significant treatment plant challenge or unanticipated change in influent wastewater flow and/or characteristics that affected treated wastewater effluent quality?
- If complacency was determined to be a contributing factor in the episode previously described, what actions have been taken to prevent it from occurring in the future? What process control procedures and special studies were completed related to the identified cause of complacency?
- How would plant staff respond to deteriorating unit process or treated wastewater effluent quality during unusual changes in influent characteristics or unit process performance?
- How would plant personnel deal with a situation in which all of the usual process control procedures do not maintain optimized performance?

Table 7-1. **Guidelines for Partnership for Clean Water: best practices for avoiding complacency**

Complacency definition: Inadequate capability (i.e., skills, policies, procedures) exists to maintain optimized treatment performance during nonroutine events and/or RPSEs.
Best practices of optimized utilities to help prevent the existence of complacency at their facilities:
• Maintaining utility awareness and commitment for responsibilities to maintain regulatory compliance and the need to consistently achieve treatment performance goals.
• Exhibiting the ability to characterize influent wastewater quality and implementing appropriate responses to significant changes detected in influent quality.
• Demonstrating an understanding of complacency (i.e., optimized plants do not take for granted their performance status nor deny the future potential for challenging situations and/or potential inability to successfully deal with challenging situations).
• Prioritizing effluent quality and compliance over short-term competing utility interests.
• Understanding the utility's potential vulnerabilities (i.e., catastrophic flooding and loss of the plant has never happened at the facility, but could it occur in the future, and how would it be addressed?).
• Embracing change and developing a formal process control program that anticipates future changes.
• Making only data driven process control decisions on the basis of established and documented procedures and not made solely on the basis of the informal, undocumented experience of the plant staff.
• Policies and procedures exist to address all potential events that could affect treated effluent wastewater quality.
• Selecting treatment technologies that are appropriate for the influent wastewater characteristics and effluent permit requirements and identifying and controlling any potential treatment risks to the best of the utility's ability.
• Effluent water quality is primarily the result of thorough planning, documented procedures, and anticipation of all potential rare and potentially serious events that may impact the treatment process, not unsustainable conservative loadings on unit processes.
• Using special studies and characterization technologies by staff is encouraged for problem solving to address specific issues that challenge the plant.

Action

If any of the best practice examples in Table 7-1 fail to exist at the utility or the potential for complacency may exist, develop an action plan to resolve and/or manage each identified cause of complacency. A number of potential actions aimed at addressing complacency are included here. This is not a comprehensive list, and utilities are encouraged to develop plant-specific actions to address any performance limiting factors identified in this area.

- Develop operational guidance for plant staff as to how to respond to significant changes in influent raw wastewater characteristics and/or compromised unit treatment process performance.
- Consider developing process control procedures and performing special studies to identify and address causes of complacency.

- Develop SOPs for staff response to deteriorating unit process performance and unusual raw wastewater quality events.
- Ensure there is a strategy for handling situations in which all of the usual process control procedures do not maintain optimized performance, including specific roles of plant personnel.
- Provide clear direction to plant staff for how to respond to situations in which the system was experiencing peak raw wastewater influent flows, but unit treatment process loading rates needed to be reduced to maintain optimized performance. Create formal policies around this area.

Training

Understanding

A robust training program is a critical component for any organization to ensure that its goals are consistently achieved and to provide an avenue to enhance the skill set of the employees. A strong message is relayed throughout the utility about its values and mission when an emphasis is placed on proper and consistent training. By investing in employee training, the company will be rewarded with a more highly skilled workforce, improved employee safety, and greater morale brought about by the attention given to the employees' professional growth.

All employees, regardless of experience, should receive regular training that covers a variety of areas, such as safety, regulatory issues, operational procedures, and emergency response. The overall program should include orientation for each new employee, refresher training for veteran employees, and procedures to evaluate the effectiveness of the training activities. Emergency response training should be included because employees need to address emergency situations with calmness and self-assurance that are the result of constant review and reinforcement of the appropriate actions.

Training should not be approached as a one-and-done event but rather a constant mindset of continuous improvement. Not all learning comes from a formal classroom setting; however, the training program should be intentional and well documented. Attention should also be given to the quality of training, as substandard efforts that are often present when the attitude is to check a box and mark a task as complete can be a waste of time and resources.

The time a utility spends in training will reap great dividends in providing effective treatment, maintaining a safe work environment, and promoting employee confidence. A comprehensive training program should be emphasized in the organization to adequately educate every employee.

Status

The status of the training program may be assessed by reviewing the following assessment questions:

- Does training/documentation exist for new employees?
  - ▲ The employee should receive training on the plant's SOPs.
  - ▲ The employee should demonstrate competency of required job skills.
  - ▲ Documentation should exist to support training activities and competencies.
- Does training/documentation exist for all employees?
  - ▲ Employee job performance should be reviewed to confirm ongoing understanding and skills to achieve the utility's operational and safety goals. This review also serves to verify that the employee is adhering to accepted procedures, and short cuts have not been developed that could jeopardize plant operations.
  - ▲ Emergency response to such items as contamination events, atypical treatment conditions, equipment failure, and other vulnerabilities should be reviewed and practiced.
  - ▲ Performance evaluation standards should be used to verify water quality testing accuracy and reliability.
  - ▲ A program to track and address continuing education requirements for operator licenses should be established.
  - ▲ A formal budget line item should be established to ensure adequate funding is allocated for employee training needs.
- Does the utility ensure that employees receive appropriate safety training?
  - ▲ Safety training should include topics required by law as well as those that are deemed as safe operating practices for the individual plant. It may be beneficial to have the employees participate in a job safety analysis for their specific job tasks.
- Is the plant staff knowledgeable about wastewater regulations and emerging technologies?
  - ▲ Regulatory information and wastewater industry publications should be readily available to the employees.
  - ▲ Informal review of wastewater treatment topics at tailgate meetings or in the course of the workday is recommended.
  - ▲ Employees regularly attend training seminars and continuing education courses to update knowledge of wastewater issues.

- Does a culture of learning and continuous improvement exist throughout all levels of staffing?
  - ▲ The utility should encourage and develop an atmosphere of learning that enables all employees to achieve their highest potential.

Action

If a formal training program is limited in nature or is not in place, action plans should be developed to improve employee knowledge and skills. Actions to consider implementing include the following:

- Development of an orientation manual for new employees
- Utilization of evaluation forms to confirm and document training activities
- Confirmation of water analysis competencies with performance evaluation standards
- Routine review of procedures with each employee to ensure consistency of tasks with all staff
- Development of tabletop scenarios and exercises to practice and evaluate emergency response
- Confirmation that all required safety training is being conducted (this may include, but is not limited to, OSHA or similar safety regulations as well as the utility's safety code)
- Provide easy access to educational materials, such as industry journals, manuals, and videos
- Participation in training classes and seminars offered by outside agencies

Staffing

Understanding

Utility staffing must be sufficient to maintain plant performance goals on a continuous basis. This is especially important at the treatment plant, where a professional operating staff is required at *all times* to make appropriate process adjustments. Proper staffing is critical with respect to the levels of process control and monitoring necessary to ensure that performance goals are achieved on a continuous basis. For a plant operating continuously, staffing levels should optimally allow 24-hours-a-day, 365-days-a-year coverage of the plant, including allowances for staff leave or attrition of workforce due to extensive retirement throughout the organization (e.g., vacations, sick leave, or training).

Some utilities use instrumentation with alarms and plant shutdown capability as a substitute for 24-hour plant coverage. Many beneficial automation tools exist for wastewater treatment plant control. It is important not to replace operator judgment and operational expertise with the control system. Refer to AWWA Standard G510 – Standard for Wastewater Treatment Plant Operation and Management, for more detailed information regarding unstaffed treatment plant operation.

Status

The status of plant staffing as it relates to optimized performance may be assessed by reviewing the following items:

- Is the number of plant staff adequate to maintain optimized performance?
  - ▲ The adequacy of plant operating staff quantity may be assessed using the following criteria and judgment:
    - ▪ Ensuring adequate coverage based on number of operators, shift length, and plant operating hours.
    - ▪ Do all treatment plant operators possess the appropriate level of certification or licenses to operate the treatment plant?
    - ▪ Utilities with staff that have accumulated significant vacation benefits may require more operators.
    - ▪ Laboratory staffing is dependent on the analyses conducted by the utility and the number of samples that are sent to contract laboratories for analysis.
    - ▪ Temporary or part-time workers, used to fill in schedules, must have sufficient skills and experience.
    - ▪ The quantity of maintenance and skilled trades staff (e.g., instrumentation or pump mechanics) is dependent on plant size, amount of mechanical equipment, and degree of instrumentation.
    - ▪ For large and medium utilities, a treatment optimization professional position and/or a separate department indicates a commitment to a proactive, multiple barrier approach to wastewater treatment.
- Is the plant routinely operated without staff present? (*Note:* This is not a desired option.)
  - ▲ Review plant staffing on weekends, holidays, and vacations for periods of unstaffed operation.
  - ▲ If the plant is operated without continuous staffing, alarms with shutdown capability are required for key parameters that may endanger public health or the environment or that may compromise compliance. Similar alarms are recommended on all processes critical to plant operation.

- Does the plant have the instrumentation to continuously monitor all critical parameters with alarms whenever the plant is operated without staff? (*Note:* Operating without staff is not a desired option.)
  - ▲ If a plant is operated without staff, alarms should be able to shut down the plant and call an operator for help within an appropriate time frame if plant conditions approach unacceptable levels.
  - ▲ If alarms are depended on to protect effluent quality, a quality control program or preventive maintenance program conducted by skilled personnel must be in place.
- Does the current staffing level have a detrimental effect on plant operation, maintenance, or laboratory procedures?
  - ▲ A sufficient number of people need to be available to perform process control activities, maintain the plant, conduct laboratory procedures, and perform other necessary operations.
  - ▲ Determine if excessive overtime or the inability to take vacations exists. Review overtime records to determine if excessive overtime is being accumulated.
  - ▲ Determine if temporary or part-time employees who are used to fill in schedules have the skills necessary to support effluent water quality and treatment goals.
  - ▲ Assess how emergency measures will be immediately implemented should a solo operator become incapacitated if the utility has only one operator per shift in a 24-hour, 365-days-per-year schedule.
- Does a high staff turnover rate, high degree of absenteeism, or large number of grievances indicate other underlying staffing problems?
- Does the improper distribution of adequate staffing result in poor plant performance?
  - ▲ The improper distribution of adequate staffing can prevent process adjustments from being made or cause them to be made at inappropriate times, resulting in unacceptable plant performance.
  - ▲ Persons responsible for process control should not be required to perform tasks not related to maintaining treatment plant performance goals (e.g., grass cutting or snow shoveling).
- Does a low pay scale or benefit package discourage more highly qualified personnel from applying for utility positions or cause personnel to leave once they are trained?
  - ▲ Pay scales and benefits should be commensurate with other utilities in the area to attract and maintain a professional staff.

Action

If staffing levels are not adequate as determined by a review of the criteria in the status section, action plans should be developed to address the identified limitations, and improvements should be implemented to optimize performance. Consider the following actions to address staffing-related performance limiting factors.

- If the plant is routinely operated without staff present and without an alarm or automatic control system, consider developing an implementation plan to staff the plant on a continuous basis (24 hours per day, 365 days per year) or set up automatic controls, including shutdowns and staff notifications, to protect effluent wastewater quality and treatment process performance.
- If instrumentation and alarms are depended on to protect effluent quality and treatment process performance in the absence of competent operating staff and a documented quality control/preventive maintenance program, develop an implementation plan to staff the plant continuously or to properly maintain a plant instrumentation and alarm system.
- If the quantity of staff available to operate and maintain the plant(s) and conduct necessary laboratory operations is insufficient and is the cause of poor plant performance, develop an implementation plan to add staff or to address the limitation through other means, such as shared personnel with other utilities.
- If the number of staff is adequate but they are distributed improperly to obtain optimized plant performance, develop an implementation plan to redistribute the workforce to support consistent achievement of plant optimization goals.

Funding

Understanding

Financial resources are required to pursue optimization as a utility goal. Adequate rate structures provide the funding for obtaining professional operating, maintenance, and laboratory staff; training of these personnel; purchasing treatment chemicals and laboratory supplies; and maintaining equipment and existing facilities necessary to efficiently and consistently produce high-quality effluent water on a continuous basis. When reviewing financial information, the impact of bonded indebtedness on the utility should be determined as should whether the rate structure creates sufficient revenue to adequately support the utility given the indebtedness.

Another important potential issue is the administrator's desire to create and maintain a self-supporting utility. Sometimes managers of small utilities create enough debt to enable the utility to be eligible for government grants. This can be especially damaging to the long-term stability of the utility and to plant performance because it allows few options for financing improvements that may be necessary to meet current and/or future wastewater regulations.

Utility funding is often maintained at levels too low to optimize performance because of the desire to avoid rate increases. Effective utility managers have the ability to overcome perceived resource barriers by properly prioritizing resource allocations so that they can be proactive in developing adequate rate structures to support the utility's optimization goals.

Status

To assess the status of utility funding as it relates to the plant achieving optimized performance, review the following items:

- Does the utility have the financial health to be self-sufficient? Do revenues cover expenses, and is there sufficient funding to cover unexpected expenditures?
  - ▲ In assessing the ability of the current rate structure to adequately support the plant, consider the following:
    - Adequate staffing to meet treated wastewater quality goals.
    - Capital improvement and replacement plans.
    - Funding to allow for adequate staff training for operators, management, and laboratory personnel.
    - Sufficient supplies of operating chemicals and laboratory equipment.
    - Adequate staffing and equipment for a preventive maintenance program.
    - Funding of a capital reserve fund in the annual operating budget.
- Review all major expenditures over the last five years and planned future expenditures. Do past and planned expenditures support the utility's optimization and efficiency goals?
- Does the utility's bond indebtedness limit funds available for other needed items?
  - ▲ Interest payments should be less than 25 percent of the utility's budget.
- Is a declining population expected to reduce anticipated revenues? Review future projections so that these situations can be accounted for in future planning.

Action

If, after a review of the criteria in the status section, the utility determines that financial resources are not adequate to support pursuing the utility's optimization goals, action plans should be developed to address the identified limitations, and improvements should be implemented.

- If there are revenue shortfalls, develop an action plan to develop a rate structure that is adequate to support all of the long-term needs of the utility. In this analysis and planning process, also include sufficient funds for unexpected expenditures or capital reserve funds.
- If there is a history of expenditures that do not support the utility's optimization goals, assess the decision-making and prioritization process that contributed to these decisions.
- If interest payments caused by an excessive amount of bonded indebtedness negatively affect the availability of funds for key utility operations, develop an action plan to reduce the amount of long-term bonded indebtedness. Also, assess the use of bonds to finance anticipated future expenditures.

Performance Limiting Factors Summary

Table 7-2 summarizes the administrative factors that may be limiting performance of the plant. Check whether the factors in Table 7-2 are Optimized and Documented, Partially Optimized, or Not Optimized. Those factors checked as Partially Optimized or Not Optimized will be prioritized in chapter 8 on identification and prioritization of performance limiting factors so that an action plan may be created for their improvement.

Table 7-2. Administration performance limiting factors summary

Self-assessment category	Questions for gauging optimization status	Optimization status		
		Optimized and Documented	Partially Optimized	Not Optimized
Administrative policies	Do the utility's overall goals and commitment at the highest level of management indicate a focus on treated wastewater effluent quality and energy efficiency and include the role of the wastewater treatment plant and its operation in achieving these goals?			
	Does the utility have a strategic planning process?			
	Does the operating staff have the authority to make required operation, maintenance, or administrative decisions?			
	Do management styles, organizational capabilities, or communication practices at any management level adversely affect performance?			
	Do administrators have a firsthand knowledge of plant needs through visits or discussions with utility staff?			
	Does the utility have ongoing public information activities, including customer outreach?			
	Has the utility maintained awareness of existing and impending regulations and has it used this information to prepare long-term plans to ensure compliance?			
	Does management support utility staff involvement in professional organizations?			
	Does management prioritize compliance over treatment costs?			
Acceptance of optimization goals	Does operational guidance reflect the optimization goals adopted by the treatment plant? • Are plant SOPs aligned to support the multi-barrier approach and plant optimization goals? • Are stated optimization goals enforced and followed even during periods when treatment is challenging or unanticipated events are encountered? • Is there provision within the plant unit process operational guidance requiring operator intervention and/or action taken prior to unit process performance exceeding the optimization goals?			

Self-assessment category	Questions for gauging optimization status	Optimization status		
		Optimized and Documented	Partially Optimized	Not Optimized
Acceptance of optimization goals *(continued)*	Do process control narratives of unit treatment processes incorporate the operational goals into the control system in the creation of safety nets? • Are alarms set up within the control system to give the operators advanced warning of flow or process loading changes? • Are there alarms set up within the control system to give the operators advanced warning of potential compliance excursions in effluent quality?			
	Does the plant operations staff maintain goals even during challenging treatment events that require reductions in production rates to meet goals?			
Involvement of all parties in the Partnership process	Has the organization developed a vision and/or mission statement that incorporates the Partnership for Clean Water philosophy?			
	Is there strong leadership throughout the organization that supports the Partnership for Clean Water culture of excellence?			
	Have all personnel been briefed on the purpose of the Partnership for Clean Water program, and do they have a good understanding of the overall intent to optimize treatment as well as the benefits of being a member of the Partnership for Clean Water?			
	Do the employees feel that their roles are well defined and do they understand their roles in the optimization process?			
	Do administrators set goals and provide an environment that expects and encourages involvement in optimization at all levels?			
Documentation demonstration of addressing complacency	Does the utility use the best practices listed in Table 7-1 to combat complacency?			

Continued

Self-assessment category	Questions for gauging optimization status	Optimization status		
		Optimized and Documented	Partially Optimized	Not Optimized
Documentation, demonstration of addressing complacency *(continued)*	During the past 10 years, how has plant staff responded to a significant treatment plant challenge or unanticipated change in influent wastewater flow and/or characteristics that affected treated wastewater effluent quality?			
	If complacency was determined to be a contributing factor in the episode previously described, what actions have been taken to prevent it from occurring in the future? What process control procedures and special studies were completed related to the identified cause of complacency?			
	How would plant staff respond to deteriorating unit process or treated wastewater effluent quality during unusual changes in influent characteristics or unit process performance?			
	How would plant personnel deal with a situation in which all of the usual process control procedures do not maintain optimized performance?			
Training	Does training/documentation exist for new employees?			
	Does training/documentation exist for all employees?			
	Does the utility ensure that employees receive appropriate safety training?			
	Is the plant staff knowledgeable about wastewater regulations and emerging technologies?			
	Does a culture of learning and continuous improvement exist throughout all levels of staffing?			
Staffing	Is the number of plant staff adequate to maintain optimized performance?			
	Is the plant routinely operated without staff present? (Note: This is not a desired option.)			
	Does the plant have the instrumentation to continuously monitor all critical parameters with alarms whenever the plant is operated without staff? (Note: Operating without staff is not a desired option.)			

Self-assessment category	Questions for gauging optimization status	Optimization status		
		Optimized and Documented	Partially Optimized	Not Optimized
Staffing *(continued)*	Does the current staffing level have a detrimental effect on plant operation, maintenance, or laboratory procedures?			
	Does a high staff turnover rate, high degree of absenteeism, or large number of grievances indicate other underlying staffing problems?			
	Does the improper distribution of adequate staffing result in poor plant performance?			
	Does a low pay scale or benefit package discourage more highly qualified personnel from applying for utility positions or cause personnel to leave once they are trained?			
Funding	Does the utility have the financial health to be self-sufficient? Do revenues cover expenses, and is there sufficient funding to cover unexpected expenditures?			
	Review all major expenditures over the last five years and planned future expenditures. Do past and planned expenditures support the utility's optimization and efficiency goals?			
	Does the utility's bond indebtedness limit funds available for other needed items?			
	Is a declining population expected to reduce anticipated revenues? Review future projections so that these situations can be accounted for in future planning.			

CHAPTER 8

IDENTIFICATION AND PRIORITIZATION OF PERFORMANCE LIMITING FACTORS/ACTION PLAN DEVELOPMENT

This chapter describes the process of identifying and prioritizing performance limiting factors so that the utility may develop an optimization action plan. All plants completing the self-assessment process should complete this chapter of the guide to prioritize factors limiting optimized performance and develop an action plan to enable improvement.

The self-assessment process has thus far gathered information about system performance and factors that could be limiting optimized performance in the areas of capacity, design, maintenance, operations, facility energy efficiency, and administration. The next part of the self-assessment is the evaluation of all the information that has been collected and the identification and prioritization of the primary factors contributing to less-than-optimized performance. This step is critical in defining the focus of activities that must be implemented to move the system toward optimized performance. For this reason, the assessment team must continue to include individuals from all departments with a vested interest in realizing plant improvements, including operators, supervisors, managers, water quality laboratory, and administration. If any of the areas for improvement include treatment changes, representatives from the treatment staff and management should also be included in the process.

This chapter (which is based on Renner et al. 1991 and Partnership for Safe Water 2011 and 2015) outlines a method to identify and prioritize factors limiting the optimized performance of a wastewater treatment plant. In addition, tools that can be used to correct the identified limitations are presented. This chapter is divided into discussions of identifying performance limiting factors, prioritizing those factors, and developing an action plan (as outlined Table 8-1) to improve performance over time.

In order to optimize its facility, a utility must first be able to identify factors that are contributing to less-than-optimized performance. Second, the list of identified factors must be prioritized, taking into account what is required to address the factors in terms of impact, urgency, cost, and time to implement a solution. The utility may want to develop two lists: one containing items that can be accomplished in the near term and the other for long-term solutions. In this way, plant staff can address these issues through solutions that are more easily implemented while simultaneously working on ones that require a more significant amount of effort, resources, and time. Third, the utility should begin putting together action plans to solve the prioritized problems. Fourth, the utility should carry out the action plans, monitoring and recording progress over time, and making adjustments to the plans as needed to ensure successful completion. Throughout this process, the utility should keep detailed records of activities completed and the outcome. Partnership for Clean Water subscriber utilities provide regular optimization activity updates as a component of the program's annual reporting process. The steps of the action plan development process are summarized in Table 8-1.

Identification of Performance Limiting Factors

Throughout this guide, the self-assessment team has been presented with a series of questions, which are designed to provide introspective thought on the part of the utility. After carefully considering a question, the utility should decide whether that part of their plant, operation, or facility is optimized, partially optimized or not optimized. Table 8-8 provides a summary list of all the self-assessment questions included in all previous chapters of this guide. These questions are listed in the order in which they appear in this guide. Note that not every self-assessment question will apply to every treatment plant. Plants are only required to address the self-assessment questions that are pertinent to their treatment plant and processes. Explanations of the applicability of specific self-assessment questions appear throughout the guide.

As part of this process, the assessment team should identify whether or not an item is truly affecting plant performance as items can be interrelated, making the root cause of the problem difficult to identify. For example, on the basis of the performance potential graph, the utility could identify that an undersized

Table 8-1. Steps to action plan development

1. Identify performance limiting factors
2. Prioritize performance limiting factors
3. Develop action assignments for the continuous improvement plan
4. Measure results of actions, make adjustments, reevaluate, and continue the optimization sequence

aeration basin is causing poor treated wastewater effluent quality without having investigated whether or not the poor water quality is a result of other factors, such as poor aeration or sludge management practices that have not been optimized. To discern the cause, a special study or series of special studies may need to be conducted with the aim of proving or disproving the suspected cause of the poor performance.

As special studies are conducted, some items that were previously of concern may be removed from the list while others may be added. An example of a special format study is included in the Appendix B of this guide. Establishing buy-in and support from all participants for the common goal of improving plant performance is critical at this stage. For example, operators that have become complacent because of a consistent raw wastewater influent quality need to recognize that complacency and become engaged in developing strategies that encourage them to remain vigilant about effluent quality and unit process performance throughout the treatment process. Similarly, and perhaps more difficult to address, are administrative issues, such as inadequate funding or lack of communication, which must be addressed with the support of plant and utility management. Some issues may be resolved quickly, and others may require significant planning and budgetary considerations.

Areas addressed during the self-assessment process represent areas in which the utility has the opportunity to improve, including performance, unit processes, operations, and administration. The detailed questions included in previous chapters of this guide help utilities to identify potential areas for improvement. Examples of the broad topics addressed in the previous chapters are outlined in the following sections. The self-assessment completion report should discuss the responses to the self-assessment questions, including any necessary documentation to support specific responses. Utility staff completing the self-assessment report should provide detailed narratives that describe the team's rationale and activities as they work to achieve treatment plant optimization. A series of tables at the end of this chapter will assist the utility in organizing its responses and formulating a performance improvement action plan.

Prioritization of Performance Limiting Factors

Once the self-assessment process has been completed, including any special studies and assembly of a final list of performance limiting factors, the utility then needs to prioritize the list to create a roadmap to optimization of their facility. Performance limiting factors should be prioritized in order of their impact on achieving optimized performance as well as their urgency. A numeric rating should be assigned to both of these categories in order to quantify the priority of the action items identified during the self-assessment process.

Prioritization of performance limiting factors is accomplished by a two-step process. First, all factors that have been identified are individually assessed with regard to their impact on optimized performance, and they are assigned a 1 through 5 rating, with 1 equating to minimal impact on performance and 5 equating to the most significant impact on optimized performance. This rating is typically performed for assessment categories receiving a Partially Optimized or Not Optimized rating. Categories that are Optimized and Documented typically do not result in action items and thus are not rated.

The second step in prioritizing performance limiting factor categories that are not fully optimized is to list them in order of urgency, assigning a rating from 1 (least urgent) to 5 (most urgent). List the highest rated Partially Optimized or Not Optimized factor categories in order of assessed urgency. The prioritized summary list of factors provides a valuable reference for the next phase of the assessment, which is developing an action plan and implementing improvements to address the performance limiting factors and move the system toward optimized performance. An example of the rating scale for prioritizing performance limiting factors is included in Table 8-2.

The numerical rating of performance limiting factors is designed to be a beneficial process for utilities to help guide plant optimization actions. However, the resulting ranking may be adjusted on the basis of internal plant or utility factors. The key to the assessment is to prioritize the *top* items affecting performance so that clarity is provided to utility personnel implementing optimization activities. In addition, the remaining factors that are not highly rated still represent a significant finding and should be addressed after the higher rated items are corrected. Lower ranking factors that are inexpensive and easy to complete should also be implemented rapidly—do not wait for completion of the self-assessment report to address these items as they may ideally be addressed and corrected at the time they are identified. These factors are also a source for providing recognition to utility personnel for adequately addressing those potential sources of problems and making rapid improvements in operations, efficiency, and treated wastewater effluent quality.

As previously mentioned, two lists should be developed: one for short-term actions and the other for longer-term, more significant actions. The rationale for this is that if the first item on the list is going to require a great deal of time and/or resources to address, the utility may also simultaneously be making incremental progress on other issues that require less time and resources to correct.

Table 8-2. Rating scale for prioritizing performance limiting factors

Rating	Description
5	Major impact on long-term optimization goals, sustained
4	Major impact on short-term optimization goals
3	Important impact on optimization
2	Minor impact but sustained
1	Minor short-term impact

For example, a utility may identify the performance of the primary clarifier solids removal system as a primary performance limiting factor during the self-assessment process. After examination of the many factors that can contribute to poor primary clarifier performance, the assessment team determined that the problem was mechanical, resulting in an action plan to replace the clarifier's solids handling and removal system. This is a long-term action that will take a lot of planning to select the proper equipment and design supporting facilities. Implementation will also need to be carefully planned to take the existing equipment out of service, remove the old equipment, install new equipment, and tie it in to SCADA control. In addition to the time involved, the utility will need to plan for funding as this is a major undertaking. If all of the attention is focused on this major task, the utility could miss opportunities to address smaller issues that cumulatively may have a major impact on operations and water quality. For example, they may identify chemical application points that would provide better treatment if they were moved or an improved means of sludge mass control. This type of activity can be carried out by the plant staff and individually accomplished with minimal planning, downtime, and expense.

Developing Action Plans

The goal of conducting the self-assessment and addressing the identified performance limiting factors is to achieve the desired level of performance for an existing wastewater treatment plant without making major modifications and capital investments. The areas in which performance limiting factors have been broadly grouped (performance, administration, unit processes, facility energy efficiency, and operation) are all important in that a factor in any one of these areas can individually cause poor performance. However, when a utility is implementing optimization activities, the relationship of these categories to achieving the goal of efficiently producing a consistently compliant treated effluent wastewater must be understood.

The prioritized performance limiting factors should be categorized into the four groups (performance, administration, operations, and unit processes) so that the resources needed to resolve the issue can be assembled. Some issues may not fit neatly into just one group and may cross over into two or more groups. A strong and detailed action plan should contain as many of the following pieces as possible:

- definition of the issue
- identification of the responsible person or department to complete the action
- required tasks
- target completion timeline
- parameters needed to measure success
- any budget requirements

The Partnership for Clean Water's Phase III Self-Assessment Completion Report Template document contains blank action plan tables, utilizing the format displayed in these examples that guide utilities to create detailed action plans for plant optimization activities. The following examples demonstrate how the resources required to complete an action tend to be group-specific. When assigning tasks to groups, do not overlook the importance of the operations staff to implementing plant improvements.

These are general examples that may not be applicable to all treatment plants. Plant staff should develop action plans that are specific to their plant and address the performance limiting factors identified during the self-assessment process.

Example 1 (Performance)

The wastewater treatment plant is not meeting its treated effluent discharge permit requirements for ammonia. The plant staff has assessed the plant's influent water quality for any changes, such as major increases in ammonia or the presence of nitrification inhibitory substances, and has not identified any significant changes in influent quality. Therefore, it was determined that the plant's nitrification process is not optimized. The plant staff tends to operate the plant on the basis of experience and how it has been operated in the past with little quantification of key control parameters. The plant's new process optimization specialist suspects that the deficiencies in nitrification performance and high effluent ammonia concentrations may be a result of poor sludge mass control. One possible solution to this issue is to implement the application of a sludge mass control program at the plant, setting targets for aeration basin and secondary clarifier sludge volumes as well as target sludge wasting rates. Sludge volumes and effluent quality can be tracked during this time to evaluate the effectiveness of the implemented performance improvement measures. These solutions do not require any outside resources or funding but rather a commitment to improve and the tenacity to find and solve problems. An example of how this improvement strategy may be expressed in an action plan format is displayed in Table 8-3.

Example 2 (Administrative)

The plant is typically staffed by one operator at a time who has no time during the day to do anything other than keep the plant running. The administration needs to recognize that this staffing situation leaves no time for troubleshooting issues, staff training, or performing special studies to improve performance, potentially resulting in compromised plant performance and effluent quality. This situation can also be very tenuous because the plant staff may become unhappy and look for employment elsewhere. This then forces the utility to hire new staff who may not understand the operation of the plant and have limited time to train. In this scenario, the administration would need to commit the financial resources to hire and train more staff to provide adequate staffing to achieve the

Table 8-3. Example action plan for improving nitrification performance

Issue	Short-term solution	Person(s) responsible	Target date to be completed	Long-term solution	Person(s) responsible	Target date to be completed
The wastewater treatment plant is not meeting discharge permit requirements for ammonia	Address issue by setting process targets for sludge mass and regularly collecting and trending effluent ammonia concentrations.	Process optimization specialist, operators	December 2016	Institute a formal sludge mass control program, in which critical sludge mass control parameters are measured on a daily basis, institute daily operator meetings.	Treatment plant supervisor, process optimization specialist, lead operator, all operators	June 2017

plant's water quality objectives. An example of how this improvement strategy may be expressed in an action plan format is displayed in Table 8-4.

Example 3 (Operations)

The staff has the desire to do a good job, but they are lacking in skills and equipment. Having properly trained, competent operators with the tools required to do their jobs effectively is the responsibility of the administration. In this case, management may need to commit the resources necessary to adequately train staff through in-house sessions, outside workshops, and equipment manufacturer representatives as new equipment is purchased. The problem may also be the result of poor priority-setting capabilities as the financial resources for training and equipment purchases exist but are not being spent where needed. An example of how this improvement strategy may be expressed in an action plan format is displayed in Table 8-5.

Example 4 (Design)

The utility has a strict equipment replacement policy, such that equipment that is still serviceable but beyond its expected life cycle is replaced at the expense of equipment that is functioning poorly, but still within its expected service life. In this instance, the utility needs to recognize that equipment replacement decisions need flexibility, such that new equipment purchases have the most significant impact possible on plant efficiency and effluent quality. The utility has

Table 8-4. Example action plan to address staffing issues

Issue	Short-term solution	Person(s) responsible	Target date to be completed	Long-term solution	Person(s) responsible	Target date to be completed
Low staffing rate resulting in poor employee morale and limited time for improving plant performance.	Hire temporary operator for relief staffing. Obtain budget approval for three new operator positions.	Utility director, plant supervisor	October 2016	Hire and train three new operators. Develop a new shift schedule that can be maintained long term.	Utility director, plant supervisor	December 2017

Table 8-5. Example action plan to address operational issues due to inadequate training

Issue	Short-term solution	Person(s) responsible	Target date to be completed	Long-term solution	Person(s) responsible	Target date to be completed
Operators do not have adequate training or equipment available to perform routine process control analyses.	Send operators to local training workshop. Budget for new online DO meter in 2017.	Plant supervisor	October 2016	Develop long-term training plan for operators. Purchase, receive training on, and develop SOPs for new online DO meter.	Plant supervisor, lead operator, all operators	April 2017

the resources necessary, but they are not being used to their fullest extent, and a change in procedures is needed. An example of how this improvement strategy may be expressed in an action plan format is displayed in Table 8-6.

Example 5 (all the above)

The plant is capable of producing a consistently compliant treated wastewater effluent, but administration places an emphasis on operational costs over quality. On the surface, this may appear to be solely an administrative issue, but it actually has roots in all four categories. The plant is not meeting optimization goals because they are overlooked at the expense of minimizing costs. Priority-setting

Table 8-6. Example action plan to address equipment replacement policy

Issue	Short-term solution	Person(s) responsible	Target date to be completed	Long-term solution	Person(s) responsible	Target date to be completed
Lack of flexibility in equipment replacement policy results in operational inefficiencies poor effluent quality.	Revise equipment replacement policy and attain board approval.	Plant manager, utility manager, board	December 2016	Reprioritize equipment replacement needs and adjust 2017 budget accordingly.	Plant manager, utility manager	June 2017

capability is not being demonstrated because the focus on cost minimization discourages plant staff from fully evaluating all potential solutions to achieving its effluent quality goals, for example, implementing changes to the plant's sludge mass control practices to better optimize sludge age, mass, and retention times in the aeration basin and secondary clarifiers. To accomplish any of the solutions in this example, the utility will need to develop short-term and long-term action plans, including strategies regarding capital investment or actions that may take more significant resources to implement. Cost may remain a factor in the selection of the implemented solution, but it should not be the primary consideration at the expense of maintaining a consistently compliant effluent. An example of an action plan to address one of these factors is displayed in Table 8-7. Note that sometimes long-term actions will be dependent on the outcome of the short-term actions that are identified as demonstrated in this table.

Implementing Action Plans

At this point in the self-assessment process, the utility should have a prioritized list of actions to complete that includes:

- identification of the issues being addressed,
- specific actions and tasks,
- required resources,
- target completion date and/or timelines, and
- responsible individual(s) or groups.

As action plans are implemented, the utility needs to periodically ask the question, "Are we headed in the right direction, and is progress occurring at the expected pace?" If the answer to either of these questions is no, the utility needs to reassess

Table 8-7. Example action plan to address long-term wastewater effluent quality improvements

Issue	Short-term solution	Person(s) responsible	Target date to be completed	Long-term solution	Person(s) responsible	Target date to be completed
Administrator emphasis on cost savings can compromise treated wastewater quality.	Investigate and recommend all potential solutions to improve effluent quality.	Utility engineer, utility director, plant manager, plant operators	April 2017	Implement or make significant progress toward implementation of recommended solution.	TBD depending on solution identified.	December 2017

the approach it is taking to solve the problem and, potentially, modify the action plan in order to achieve the desired outcome. The change could be relatively minor (such as a modification of the target completion date) or very significant, based on evaluation of the goals, progress, and timeline. Utilities and plants can and do change, and the action plans developed during the self-assessment process should change with them as appropriate. Establishment of periodic check-in points for action plans are critical because they provide opportunities for critical evaluation of the progress being made. At these points, individuals should be involved in the progress assessment who are objective and can provide constructive criticism when appropriate.

When implementing action plans, some action items can be a long-term process, requiring a long-term commitment from the utility. The rationale for this is described here.

- *Greater effectiveness of training to accomplish skills transfer:* Operator and administrator training can be conducted under a variety of actual operating conditions. This approach allows transfer of the skills necessary to maintain optimized and efficient performance even during periods when the plant is under stress. Training should be conducted over an extended period to encounter many operating conditions and situations.
- *Time required to make administrative changes:* Administrative changes often require an extended period of time to address. For example, if the utility rate structure is inadequate to support optimized treatment plant performance, extensive time can be spent developing required changes in the rate structure and gaining political support to implement them. Communication barriers between labor and management may need to be addressed for improved performance.
- *Time required for minor facility modifications or to implement deferred maintenance activities:* For changes requiring financial expenditures, both

time and a multiple-step approach are typically required to gain administrative approval. First, the need for minor modifications or deferred maintenance improvements must be demonstrated through plant control efforts. Then, the administrators must be shown the need and ultimately convinced to approve funds necessary for the improvements. These activities can take considerable time before the identified modification or deferred maintenance item can be approved and corrected.

Experience has shown that no single approach can address the unique combination of factors in every wastewater treatment plant. Therefore, the actual details of implementation are plant-specific and should be completed by utility personnel. Tools to aid in action plan development and implementation, such as a special study and SOPs are described in greater detail in the appendix.

The optimization process is a continuous process as utilities are encouraged to measure progress, reassess performance, and develop new and relevant optimization actions on an ongoing basis. This process is reinforced through the Partnership for Clean Water's annual reporting process.

Conclusion

Congratulations on taking the first important step on the journey to wastewater treatment plant optimization: performing a self-assessment! You have undertaken a critical evaluation of your plant and how it operates. You have identified challenges that are limiting performance and the ability to achieve optimized performance. Now the work begins.

Refer to the action plan tables developed from the self-assessment findings from your plant and begin the journey. The value of the self-assessment will continue to reveal itself as operations team members continue to pursue treatment excellence in the days, weeks, and years to come. The strength of the self-assessment process lies in understanding the details that enable and empower staff to efficiently and consistently produce a treated wastewater effluent that meets or exceeds all applicable regulatory requirements. As Water Research Foundation Executive Director and one of the authors of the original *Self-Assessment Guide for Surface Water Treatment Plant Optimization,* Rob Renner, said, "Operational excellence is doing simple things exceedingly well."

The Partnership for Clean Water challenges operators who have participated in a self-assessment to be deliberate at moving to the next step of optimization: addressing the performance limiting factors identified in the self-assessment and being tenacious at completing the action plans developed. As operators and maintenance team members complete action plans and performance improvements are realized, momentum will build and a culture of excellence will begin to permeate the utility. At this stage, it is critical to fan the flame and reinforce the positive steps completed while continuing to encourage and support future plans for improvement. The journey to operational excellence will require tenacity, patience, perseverance, communication, and the commitment of all operations team members. The journey is long, and the end goal of excellence in optimization is truly never reached; there will always be additional improvements that can be implemented and additional performance gains that can be achieved. In the end, the lessons learned during the journey of striving for excellence are the reason for taking the first step.

Table 8-8 is a summary of the self-assessment questions that appeared throughout the document, arranged by chapter and topic.

References

Partnership for Safe Water. 2011. *Self-Assessment Guide for Distribution System Optimization (Working Draft)*. Denver, CO: AWWA.

Renner, R.C., B.A. Hegg, J.H. Bender, and E.M. Bissonette. 1991. *Handbook—Optimizing Water Treatment Plant Performance Using the Composite Correction Program*. EPA 625/6-91/027. Cincinnati, OH: USEPA.

Table 8-8. Self-assessment question summary

Chapter 2. Performance Assessment				
		Optimization status (choose one)		
Self-assessment category	Questions for gauging whether system is optimized	Optimized and Documented	Partially Optimized	Not Optimized
Performance assessment	Did the final effluent quality meet the PCW (Table 1-1) or plant-specific performance goals?			
	Do the primary clarifiers achieve 25 to 40 percent removal of BOD_5?			
	Do the primary clarifiers achieve 50 to 70 percent removal of total suspended solids (TSS)?			
	If the plant has multiple process trains, does each train perform equally, or are there significant differences in the performance of individual process trains?			
	Do changes in wastewater flow rate, implemented to perform routine operational or maintenance practices, or influent quality impact the performance of any unit treatment process?			
	Do all of the unit processes meet their performance goals or does treated effluent quality degrade despite consistent and stable process control?			
	Are the treated wastewater effluent permit requirements being achieved even through other upstream unit processes do not meet the specified optimization goals?			
	Are there any positive or negative trends in the data that should be addressed or further evaluated?			
	Is the activated sludge barrier adequate to achieve the performance goals and scientifically calculated based on flow rate and organic loading?			
	Does the actual sludge mass produced agree with the projected sludge mass produced to within +/–15 percent?			

Continued

Chapter 3. Capacity Assessment				
		Response (select one per category)		
Unit process	Questions for addressing adequacy of system capacity	Optimized and Documented (or Type 1)	Partially Optimized (or Type 2)	Not Optimized (or Type 3)
General	Is the capacity of any major unit process insufficient for meeting the plant's average flow requirements?			
Preliminary treatment	Does the equalization basin volume provide adequate detention time to meet the plant's treatment goals?			
	Is the plant's loading compromising its capacity and ability to optimize?			
Primary treatment	Do limitations in primary clarifier design (volume or surface overflow rate) cause poor settling that results in inadequate treatment?			
	Do limitations in sludge handling and removal equipment prevent optimized primary clarifier performance?			
Secondary treatment/ sludge handling (address all that apply)	Do capacity limitations in aeration basin design limit optimized performance of the activated sludge process?			
	Do capacity limitations in SBR design limit optimized performance of the process?			
	Do capacity limitations in attached growth design limit optimized performance of the process?			
	Do sludge-handling pumps have sufficient capacity rating to handle the total solids volume, including with the largest pump out of service?			
	Is pump efficiency routinely tested to quantify performance?			
	Are there any maintenance issues limiting the performance of sludge-handling/ removal systems in use at the plant?			
	Do limitations in secondary clarifier design (volume or surface overflow rate) cause poor setting that results in inadequate treatment?			
	Do limitations in sludge handling and removal equipment associated with the secondary clarifiers prevent optimized clarifier performance?			

Tertiary treatment (address all that apply)	Do limitations in filter capacity limit the optimized performance of the process?			
	Do any limitations in backwash system design or capacity exist that limit optimized process performance?			
	Is membrane bioreactor (MBR) capacity limiting optimized performance of the system?			
Disinfection	Is the capacity of the chlorine disinfection system suitable to provide adequate disinfection?			
	Is the capacity of the dechlorination system suitable to provide an adequate amount of dechlorination?			
	Is the capacity of the UV system sufficient to meet the wastewater treatment plant's permit requirements?			
Other Processes	Does the capacity of any other unit process applied at the plant limit optimized plant performance?			
Chapter 4. Unit Process Performance Assessment				
Self-assessment category	**Questions for gauging optimization status**	**Optimization status (choose one)**		
		Optimized and Documented	**Partially Optimized**	**Not Optimized**
Source water/ influent	Have potential risks to source water/influent quality been identified?			
	Is source water/influent routinely sampled and monitored for changes in water quality?			
	Has a source water control program been implemented in the community in the form of a sewer-use bylaw or similar regulation?			
	Are industries that discharge potentially harmful or toxic substances required to pretreat their water prior to discharge?			
	Are industries required to submit regular testing and/or maintenance records to verify and demonstrate the performance of any onsite pretreatment systems?			
	Is a protocol in place to respond to source water/influent emergencies?			
Preliminary treatment	Has a regular maintenance and inspection program been developed for the facility that includes the preliminary treatment equipment?			
	Does the plant maintain an inventory of spare parts to minimize equipment downtime?			
	If the equipment is designed in a duty/ standby configuration, is there a policy to ensure equal run-times on each unit?			

Continued

Preliminary treatment *(continued)*	Has the plant optimized the source water control program to minimize fat, oil, and grease material from being discharged into the collection system?			
Equalization	Has a low operating level target been defined for the equalization tank?			
	Does the tank effluent flow method (pump versus gravity) and control strategy allow flexibility for consistent flow to be provided to the biological system even under peak flow events?			
	Has a routine inspection program been developed?			
	Have SOPs been developed for the equalization tank, including an emergency response plan?			
Primary clarification	Have primary clarifier removal targets (high and low range) been defined?			
	Does the plant sampling and testing program provide the required data to fully analyze the status of the primary clarifiers?			
	Have the appropriate trend charts been developed to visualize the performance of the primary clarifiers over time?			
	Has the operator collected and analyzed primary clarifier flow data?			
	Does the primary clarifier have a regular solids wasting program, and does the operator monitor key settled solids wasting parameters?			
	Has the primary clarifier(s) been inspected for any visual signs or indicators of other potential issues?			
	Have SOPs been developed for the primary clarification process, including an emergency response plan and maintenance procedures?			
Process energy considerations—Equalization Basin	Is the mixing intensity for the equalization tank controlled based on tank level?			
	Do the equalization tank effluent pumps operate at or near the BEP of the pump curves on a consistent basis?			
Suspended growth (CAS)	Have performance targets been developed for the secondary suspended growth treatment process?			
	Does the plant's monitoring and testing program allow for sufficient data collection to thoroughly monitor actual plant performance compared with the defined performance targets?			

Suspended growth (CAS) *(continued)*	Are the appropriate trend charts being developed and populated to capture all of the raw data and calculated performance parameters to properly monitor the aeration tank?			
	Is the plant equipped with an appropriate chemical dosing system or other means to combat foaming or filamentous events?			
	Has a regular maintenance and inspection program been developed for the plant that includes daily inspection of the aeration tank and associated equipment?			
	Have SOPs been developed for the aeration tank, including an emergency response plan?			
Suspended growth (SBR specific)	Have the appropriate cycle and phase durations been identified and optimized for actual plant operation?			
	If the plant has a nonfixed style decanter (i.e., floating or adjustable) has the operation been optimized?			
	Has the batch volume per cycle been optimized to suit the actual operation and number of cycles per tank?			
Attached growth	Have performance targets been defined for the process?			
	Has a sampling and testing program been developed for the attached growth process that allows actual performance to be monitored compared to the defined targets?			
	Has the collected data been plotted on trend charts along with any applicable performance targets?			
	Does the plant have the capability to provide sufficient aeration to meet the treatment requirements?			
	Does the attached growth process require regular replacement of the media and does the plant have sufficient spare media on site?			
	Has a regular inspection and maintenance program been developed for the attached growth process?			
	Have SOPs been developed for the attached growth process, including an emergency response plan?			
Secondary clarification	Have performance targets been developed for the secondary clarifier?			
	Has a sampling, testing, and monitoring program been put into place for the secondary clarifier?			
	Has a sampling and monitoring program been developed to monitor the settling characteristics of the sludge?			

Continued

Secondary clarification *(continued)*	Have the appropriate trend charts been developed to properly track the collected data and monitor the clarifier performance relative to the defined operating targets?			
	Has the operation of the RAS system been developed to optimize the distribution of biomass between the aeration tank(s) and the secondary clarifier and satisfy the goals of the total sludge mass control concept?			
	Has the sludge-wasting program been developed on the basis of the total sludge mass control concept that includes operating at a target SRT?			
	Have steps been taken to minimize filamentous bacterial growth that leads to poor settling?			
	Does the operator have the ability to dose chlorine for control of filamentous bacterial growth?			
	Has a maintenance and inspection program been developed for the secondary clarifier?			
	Have SOPs been developed and documented for the secondary clarifiers, including an emergency response plan?			
Process energy considerations: secondary treatment	Is the current aeration diffuser system providing acceptable oxygen transfer efficiency?			
	Does the plant use dissolved oxygen monitoring for blower speed control?			
	Is the current aeration blower technology providing the best efficiency?			
	Do the RAS pumps have speed control and is the operation tied to the raw wastewater influent flow rate?			
	If the plant has dedicated WAS pumps, do they operate consistently as close to the BEP as possible?			
	Has plant/utility staff evaluated potential upgrades for increased energy efficiency?			
Biological nitrification	Have performance targets been developed for the nitrification process?			
	Has a sampling, testing, and monitoring program been put into place that includes the required data for nitrification?			
	Have the appropriate trend charts been developed to collect the required data and track actual system performance relative to the defined operating targets?			
	Has a maintenance and inspection log been developed for the plant that includes the secondary treatment system and nitrification process?			

Biological nitrification *(continued)*	Have SOPs been developed for the nitrification process, including an emergency response plan?			
Biological denitrification	Have performance targets been developed for the biological denitrification system?			
	Has a sampling and testing program been developed that includes the denitrification process?			
	Have all data been plotted on trend charts along with applicable performance targets?			
	Has the nitrification process been optimized in accordance with the sludge mass control concept?			
	Has a nitrate mass balance been completed for the pre-anoxic denitrification process to ensure sufficient recycle of nitrate from the aeration tank?			
	If the plant includes a post-anoxic denitrification step, has the pre-anoxic tank been optimized for maximum nitrate removal? Has the appropriate testing and data collection been implemented to verify nitrate removal in each anoxic zone?			
	If post-anoxic denitrification is used, has the supplemental carbon dosing system been optimized for chemical usage efficiency?			
	If methanol is used for supplemental carbon dosing, have alternative, less hazardous chemical options been explored?			
	Has a routine maintenance and inspection program been developed that includes the denitrification process?			
	Have SOPs been developed for the denitrification process, including an emergency response plan?			
Biological phosphorus removal	Have performance targets been defined for the biological phosphorus removal process?			
	Has a total mass control program been implemented to define a target sludge mass and SRT that balances the requirements for biological phosphorus removal with total nitrogen removal?			
	Has a sampling, testing, and monitoring program been developed for the biological phosphorus removal process?			
	Has the plant optimized the acetate chemical dosing system to maximize phosphorus removal and maintain a consistent mass of phosphorus accumulating bacteria?			
	Does the plant have the capability to supplement a chemical coagulant to enhance phosphorus removal? If so, has this system been optimized according to the chemical phosphorus removal section of this guide?			

Continued

Biological phosphorus removal *(continued)*	Has the final effluent solids removal process been optimized in accordance with the applicable section(s) of this guide?			
	Have SOPs been developed for the biological phosphorus removal process that include an emergency response plan?			
Chemical phosphorus removal	Have performance targets been defined for the system?			
	Does the plant have a regular sampling, testing, and monitoring program in place that includes regular testing of the phosphorus removal system performance?			
	Do the plant operators conduct regular jar testing using actual wastewater and coagulant samples to verify the chemical dosing system set-points?			
	Has the final suspended solids removal system been optimized in accordance with the applicable sections of this guide?			
	Does the plant's control system allow for flow-pacing of the coagulant chemical to match the wastewater flow?			
	Have the plant operations staff conducted a review of potential alternate chemicals to determine the optimal coagulant?			
	Does the chemical storage and dosing system include proper containment for spills to protect the plant staff and the environment?			
	Does the plant have sufficient chemical storage volume to allow for sustained operation in the event of delayed chemical delivery?			
	Have SOPs been developed for the chemical phosphorus removal system that include an emergency response plan?			
Chlorine disinfection	Have performance targets been defined for the chlorine disinfection system?			
	Has a defined CT value been set for the plant to achieve the required level of disinfection?			
	Has a regular sampling and testing program been implemented at the plant that includes analyzing the final effluent for disinfection system performance?			
	Does the plant's control system allow for flow-paced chlorine dosing to allow for reduced chemical usage at lower flow rates?			
	Does the plant have instrumentation that allows for automated control of the dechlorination chemical dosing pumps?			
	Has the plant reviewed the various chlorine-based chemical options available for disinfection?			

Chlorine disinfection *(continued)*	Has the plant reviewed nonchlorine based chemicals as alternatives to chlorine?			
	Has a regular maintenance and inspection program been developed for the plant that includes the chlorine disinfection system?			
	Does the plant have sufficient chemical storage to deal with minor delays or disruptions to delivery of new chemical?			
	Have SOPs been developed for the chlorine disinfection system including an emergency response plan?			
UV disinfection	Have performance targets been developed for the UV disinfection system?			
	Does the operator have the capability to monitor key operating parameters for the UV disinfection system including UV intensity and effluent UV transmittance?			
	Have trend charts been developed for the UV system to monitor key operating and performance parameters?			
	Has an inspection and logging program been developed for the plant that includes the UV disinfection system?			
	Have SOPs been developed for the UV disinfection system, including maintenance procedures and an emergency response plan?			
Process energy considerations (disinfection)	For chemical disinfection (chlorine and nonchlorine), is the chemical dosage rate automatically adjusted to meet the demand?			
	For UV disinfection, does the system have the capability to vary the number of lamps in service based on UV transmittance or flow or level in the UV channel?			
Tertiary filtration	Have performance targets been defined for the tertiary filtration process?			
	Has a sampling and testing program been implemented for the tertiary filtration process that allows actual performance to be tracked and compared to the defined targets?			
	If chemical phosphorus removal is included at the plant, has the chemical coagulant dosage been optimized for use with the tertiary filter?			
	Is there a provision for continuous operation during filter cleaning and/or backwashing operation?			
	Have backwash initiation and return-to-service parameters been defined?			
	Does the control system have the ability to automatically initiate a backwash?			

Continued

Tertiary filtration *(continued)*	Has a regular maintenance and inspection program been developed for the tertiary filter?			
	Have SOPs been developed for the tertiary filter that include an emergency response plan?			
Membrane bioreactor process	Have performance targets been defined for the process?			
	Has a sampling and testing program been developed for the membrane process that allows tracking of actual system performance compared with the defined targets?			
	Has the total sludge mass control program been adapted and applied for the membrane bioreactor process?			
	Does the plant control system actively monitor the membrane permeability?			
	Does the system include provision for restoring the permeability of the membranes when they have become fouled?			
	If chemical cleaning is used, has a cleaning frequency and protocol been established?			
	Does plant have an effective method of controlling foam events?			
	Have SOPs been developed for the process that include an emergency response plan?			
Sludge dewatering	Does the plant have an appropriate method for handling any unsatisfactory polymerized sludge generated during equipment startup routines?			
	Has a sampling, testing, and monitoring program been developed for the plant that includes recycle wastewater from the solids treatment process?			
	Has the polymer dosing system been optimized to meet the requirements of the dewatering operation while minimizing excess polymer?			
	Does the polymer dosing system operation include the use of jar testing for each new batch of sludge processed?			
	Does the sampling and testing program include regular sampling of material from outside sources (i.e., sludge and septage) if applicable?			
	Have SOPs been developed for the solids handling process, including an emergency response plan?			

Chapter 5. Facility Energy Optimization Assessment				
Self-assessment category	**Questions for gauging optimization status**	**Optimization status (choose one)**		
		Optimized and Documented	**Partially Optimized**	**Not Optimized**
Building energy performance assessment	Does the wastewater treatment facility currently quantify energy consumption using energy monitoring, measurement, and verification programs and portfolio management tools that provide this information?			
	Has the total number of buildings on the WWTP site been quantified?			
	Have all of the different fuel sources used by the facility been quantified?			
	Is process energy consumption metered separately from the energy consumed by building systems?			
	Is the appropriate information (e.g., utility bills) currently compiled in a format that allows utility staff to complete this exercise?			
Energy audit	Has the utility completed an energy audit for either a single building or the entire facility?			
	Has the utility staff defined the needs and scope of work related to energy audits and efficiency upgrades?			
	Is the utility familiar with local codes and bylaws that affect implementation of energy-efficiency upgrades at the facility?			
	Are any major building renovations planned at the facility? If so, are there energy-efficient upgrades that would be beneficial to implement during these planned renovations?			
	Are utility staff aware of programs applicable to the building or project that would provide financial incentives to improve energy efficiency? If so, are utility staff aware of the rules and procedures required for eligibility to receive financial incentives?			
	Have subject matter experts been identified to assist the utility in moving project plans forward in this area?			
Re-/Retro-commissioning	Does the utility have a regularly occurring, comprehensive operations and maintenance plan that includes detailed verification of building systems function at the facility?			
	Is the facility equipped to automatically monitor the status and trends of the building systems (by a building automation system or other), and is the data produced regularly reviewed by personnel?			

Continued

Energy management	Is there an active energy manager onsite?			
	Is a team responsible for leading energy management activities onsite (i.e., a green team or similar group)?			
	Does the utility participate in any local or citywide energy management initiatives?			
	If these items do not exist, is there any staff member with the potential to be an energy manager?			
	If these items do not exist, can a consultant potentially be engaged to provide this service?			
Energy measurement and verification	Is the energy consumption of the facility measured and verified?			
	Is there staff available to monitor this data, summarize the results, and communicate them to utility personnel?			
Building and energy codes and standards	Does the plant's jurisdiction have an energy code or guideline?			
	Does the facility include buildings that were completed prior to the adoption of the current set of codes in effect in the area?			
	If deficiencies exist, has a plan been developed to bring site facilities into compliance with applicable codes and/or standards?			
Third-party certifications	Are buildings on the WWTP site full of old appliances and equipment? If so, has a plan been developed to replace inefficient appliances and equipment over time?			
	Has the utility adopted a goal that includes third-party certification at any level?			
	Does the performance of the WWTP's buildings meet the thresholds of third-party certification for energy performance?			
	Has the plant or utility adopted a policy such that new product purchases in eligible categories are rated by Energy Star, EnerGuide, or another appropriate entity?			

Chapter 6. Application of Operational Concepts

Self-assessment category	Questions for gauging optimization status	Optimization status (choose one)		
		Optimized and Documented	**Partially Optimized**	**Not Optimized**
Process control	Have performance goals been established for each treatment process in the plant?			
	Has a process control sampling and testing schedule been developed and implemented?			
	Have monitoring spreadsheets or similar data collection tools been developed and implemented that allow data to be collected and interpreted?			

Process control *(continued)*	Has an emergency response procedure been developed for loss of disinfection chemical feeds or for unacceptable plant effluent treated wastewater quality?			
	Has a comprehensive activated sludge process control strategy been developed, documented, and implemented?			
	Has all of the plant staff been involved in the development of the process control program, and have they developed operational guidelines, including emergency response guidelines?			
Operator application of concepts	Does all operating staff have the following characteristics? • The tenacity to achieve the plant's performance goals • A willingness to take responsibility for plant performance • A willingness to learn • Confidence to make changes in treatment • A tenacity to investigate process upset conditions and the confidence to implement change in an existing practice to optimize performance • Empowerment to make changes in treatment • An understanding of when to call for help and whom to call.			
	Can the plant staff conduct the necessary calculations or have easy access to computer programs or other tools to determine the appropriate chemical feed rates (e.g., mL/min, L/s, or gpm) necessary to deliver the required chemical dosage throughout the plant?			
	Can plant staff respond appropriately to changes in influent quality?			
	Does the plant staff have the knowledge and skills to repair and/or maintain process components (pumps or valves, for example) and understand the benefits of and tasks associated with a preventative maintenance program?			
	If proprietary processes are in use at the plant, has the plant staff worked with the manufacturer to determine the most appropriate controls to identify and address process upsets?			
Communication	Has a formal communication protocol been developed?			
	Do operators effectively communicate with each other?			

Continued

Communication *(continued)*	Do operators and maintenance workers effectively communicate to ensure plant maintenance status is provided to the operators frequently and accurately?			
	Is information communication encouraged between plant staff?			
	Does the utility ensure that employees understand the information presented?			
Online instrumentation, PCS, and SCADA	Has a maintenance and calibration schedule been implemented for all instrumentation?			
	Has the utility implemented a training program to train operators in the care and maintenance of on-line instrumentation? Has the utility implemented a training program to train operators in the use and operation of the PCS and SCADA systems?			
	Does the utility have clearly defined roles for who is permitted to modify and maintain the PCS and are those individuals properly trained?			
	Application and placement: Are the instruments selected suitable for the application and physical placement for which they are installed? Does the placement of the instruments allow for ease of access for maintenance? Are environmental conditions suitable for analyzer placement? Is the placement of the sample collection and delivery lines suitable to obtain a representative sample?			
	Calibration and verification: Are the instruments calibrated and verified regularly? Do operators follow the manufacturer's instructions for calibration and verification? Are calibration and verification performed at a frequency that, at minimum, meets regulatory requirements? Are calibration records maintained and accessible when needed? Has an SOP been developed for the calibration and maintenance of all critical instrumentation?			
	Are control systems tested and checked regularly to ensure proper operation and functionality when needed? This may include testing functionality of alarms and control capabilities.			
	Redundancy: Are redundant instruments and equipment installed where appropriate? Does the plant have a plan established for the procurement of all critical parts to move toward making the plant bulletproof?			
	Does the utility have provision for a backup server or control system in the event of communication loss or failure?			

Self-assessment category	Questions for gauging optimization status	Optimized and Documented	Partially Optimized	Not Optimized
Online instrumentation, PCS, and SCADA *(continued)*	Is all pertinent information documented? Is record-keeping adequate; i.e., is the necessary information maintained for the amount of time required by regulatory parties or internal plant processes?			
	Does the control system provide operational safety nets, consistent with plant practices and process performance goals, to protect the operator and treated wastewater quality?			

Chapter 7. Administrative Assessment				
		Optimization status (choose one)		
Self-assessment category	**Questions for gauging optimization status**	**Optimized and Documented**	**Partially Optimized**	**Not Optimized**
Administrative policies	Do the utility's overall goals and commitment at the highest level of management indicate a focus on treated wastewater effluent quality and energy efficiency and include the role of the wastewater treatment plant and its operation in achieving these goals?			
	Does the utility have a strategic planning process?			
	Does the operating staff have the authority to make required operation, maintenance, or administrative decisions?			
	Do management styles, organizational capabilities, or communication practices at any management level adversely affect performance?			
	Do administrators have a firsthand knowledge of plant needs through visits or discussions with utility staff?			
	Does the utility have ongoing public information activities, including customer outreach?			
	Has the utility maintained awareness of existing and impending regulations, and has it used this information to prepare long-term plans to ensure compliance?			
	Does management support utility staff involvement in professional organizations?			
	Does management prioritize compliance over treatment costs?			
Acceptance of optimization goals	Does operational guidance reflect the optimization goals adopted by the treatment plant? • Are plant SOPs aligned to support the multibarrier approach and plant optimization goals?			

Continued

Acceptance of optimization goals *(continued)*	• Are stated optimization goals enforced and followed even during periods when treatment is challenging or unanticipated events are encountered? • Is there provision within the plant unit process operational guidance requiring operator intervention and/or action taken prior to unit process performance exceeding the optimization goals?			
	Do process control narratives of unit treatment processes incorporate the operational goals into the control system in the creation of safety nets? • Are alarms set up within the control system to give the operators advanced warning of flow or process loading changes? • Are there alarms set up within the control system to give the operators advanced warning of potential compliance excursions in effluent quality?			
	Does the plant operations staff maintain goals even during challenging treatment events that require reductions in production rates to meet goals?			
Involvement of all parties in the partnership process	Has the organization developed a vision and/or mission statement that incorporates the Partnership for Clean Water philosophy?			
	Is there strong leadership throughout the organization that supports the Partnership for Clean Water culture of excellence?			
	Have all personnel been briefed on the purpose of the Partnership for Clean Water program, and do they have a good understanding of the overall intent to optimize treatment as well as the benefits of being a member of the Partnership for Clean Water?			
	Do employees feel that their roles are well defined, and do they understand their roles in the optimization process?			
	Do administrators set goals and provide an environment that expects and encourages involvement in optimization at all levels?			
Documentation/demonstration of addressing complacency	Does the utility use the best practices listed in Table 7-1 to combat complacency?			
	During the past 10 years, how has plant staff responded to a significant treatment plant challenge or unanticipated change in influent wastewater flow and/or characteristics that affected treated wastewater effluent quality?			

Documentation/demonstration of addressing complacency *(continued)*	If complacency was determined to be a contributing factor in the episode previously described, what actions have been taken to prevent it from occurring in the future? What process control procedures and special studies were completed related to the identified cause of complacency?			
	How would plant staff respond to deteriorating unit process or treated wastewater effluent quality during unusual changes in influent characteristics or unit process performance?			
	How would plant personnel deal with a situation in which all of the usual process control procedures do not maintain optimized performance?			
Training	Do training/documentation exist for new employees?			
	Do training/documentation exist for all employees?			
	Does the utility ensure that employees receive appropriate safety training?			
	Is the plant staff knowledgeable about wastewater regulations and emerging technologies?			
	Does a culture of learning and continuous improvement exist throughout all levels of staffing?			
Staffing	Is the number of plant staff adequate to maintain optimized performance?			
	Is the plant routinely operated without staff present? (*Note*: This is not a desired option.)			
	Does the plant have the instrumentation to continuously monitor all critical parameters with alarms whenever the plant is operated without staff? (*Note*: Operating without staff is not a desired option.)			
	Does the current staffing level have a detrimental effect on plant operation, maintenance, or laboratory procedures?			
	Does a high staff turnover rate, high degree of absenteeism, or large number of grievances indicate other underlying staffing problems?			
	Does the improper distribution of adequate staffing result in poor plant performance?			
	Does a low pay scale or benefit package discourage more highly qualified personnel from applying for utility positions or cause personnel to leave once they are trained?			

Continued

Funding	Does the utility have the financial health to be self-sufficient? Do revenues cover expenses, and is there sufficient funding to cover unexpected expenditures?			
	Review all major expenditures over the last five years and planned future expenditures. Do past and planned expenditures support the utility's optimization and efficiency goals?			
	Does the utility's bond indebtedness limit funds available for other needed items?			
	Is a declining population expected to reduce anticipated revenues? Review future projections so that these situations can be accounted for in future planning.			

Table 8-9. Definitions of related terms

Term	Definition
Plant type	Brief but specific description of type of unit processes employed by the plant (e.g., primary, secondary, or tertiary treatment as well as the specific types of treatment applied at the plant, such as conventional activated sludge or SBR). The plant processes will determine the self-assessment questions completed by plant staff.
Influent description	Brief description of the type of influent water received by the plant, such as municipal or industrial. If the influent received has significant industrial influence, describe the type of industry and any applicable local sewer use ordinances.
Plant performance summary	Brief description of plant performance as related to desired treated wastewater effluent quality (Partnership for Clean Water or internally established performance goals)
Optimized and Documented	Self-assessment status selected when a particular optimization category is considered to be completely optimized and documentation exists to support that response.
Partially Optimized	Self-assessment status selected when a particular optimization category is considered to be partially optimized. Limited documentation may be available to support this response.
Not Optimized	Self-assessment status selected when a particular optimization category is not optimized. Limited or no documentation may be available to support this response.
Ranking	Ordering of causes of decreased plant performance and reliability with the most critical ones listed first on the basis of impact on optimization and urgency (ranked on a scale of 1 to 5)
Performance limiting factors/category	Items identified during the self-assessment that limit plant performance. Identify factor topic category, for example, chemical phosphorus removal (unit process performance).
Action plan	A list of specific optimization actions that are planned to address the plant's performance limiting factors with the goal of achieving improvement.

APPENDIX A

STANDARD OPERATING PROCEDURE FOR USING A DESKTOP CENTRIFUGE TO PROJECT TSS CONCENTRATION

The measurement of mixed liquor suspended solids (MLSS) concentration is a critical parameter for sludge mass control in the aeration basin and secondary clarifier and is a key control parameter for the sludge mass control program described previously in this guide. Obtaining a true MLSS measurement is typically performed gravimetrically, by measuring the mass of solids retained on a glass fiber filter paper after filtration of an aliquot of sample. This MLSS procedure involves time and laboratory equipment, which may not be available to all treatment plants. The procedure outlined in this appendix provides information about how a simple desktop centrifuge test may be used to approximate MLSS concentrations for the purpose of sludge mass control. This procedure represents only an approximation of MLSS concentration for process control purposes. It is not approved for regulatory reporting. To assess the approximate accuracy of MLSS results obtained using this test, utilities should establish a correlation between the desktop centrifuge results and those obtained through traditional MLSS analytical methods, taking care to watch for potential changes in the correlation based on changes in wastewater and solids composition.

Purpose

The procedure for using a desktop centrifuge to project MLSS concentration is described here.

Scope

The standard method for measuring the MLSS concentration of a liquid involves filtering a known aliquot through a glass fiber filter paper. The filter paper is weighed before filtering the MLSS. After filtering, the filter paper is dried in an oven for 1 hour at 103°C. The suspended solids are calculated using a standard formula (see laboratory methodology in *Standard Methods for the Examination of Water and Wastewater*, part 2, page 56).

Depending on the size and configuration of the WWTP, the TSS concentration of multiple samples must be measured and interpreted each day. To collect all of the samples and complete the analysis using the Standard Methods procedure requires up to three hours per day. An effective and time-saving alternative is to use a desktop centrifuge, shown in Figure A-1, to project the suspended solids concentration of the samples. The alternative method requires only 30 minutes to collect and spin the samples in a centrifuge vial.

Approach

Calibration of the centrifuge methodology is achieved using a centrifuge spin ratio. A sample of MLSS is tested for TSS using the Standard Methods filter paper test procedure. The same MLSS sample is poured into a centrifuge vial and spun at 3,000 rpm for 10 minutes. While spinning, the centrifugal forces separate the solids in suspension from the clear water. The suspended solids compact at the bottom of the centrifuge vial. After 10 minutes, the centrifuge is stopped and the percentage of compacted solids on the graduated centrifuge vial is recorded. The percentage of compacted solids is used to calculate a centrifuge spin ratio as explained later in this section.

Responsibility

This procedure is applicable to all applicable treatment plant personnel.

Procedure for Calculating the Centrifuge Spin Ratio Using the Standard MLSS TSS Concentration

Collect a representative sample of mixed liquor from the aeration basin. In the laboratory, shake and pour the sample into a centrifuge vial. Fill the vial up to the

(Source: Gerard Wheeler, Global Facilitation, Inc.)

Figure A-1. Desktop centrifuge used for MLSS approximation test

100 percent mark on the graduated centrifuge vial. Place the graduated centrifuge vial in the centrifuge placeholder. Fill a second centrifuge vial and place it in the opposite placeholder of the centrifuge head. This will ensure the centrifuge is balanced when the rotor head is spinning (see Figure A-2). Equipment damage may occur if the centrifuge is operated in an unbalanced state. Spin the vials at 3,000 rpm for 10 minutes.

When the centrifuge rotor has fully stopped, remove the centrifuge vial from the placeholder. Read and record the percentage of solids compacted at the bottom of the centrifuge vial.

Using the same MLSS sample, perform a TSS analysis using the Standard Methods procedure. The standard method for measuring the mixed liquor suspended solids concentration of a liquid involves filtering a known aliquot through a glass fiber filter paper. The filter paper is weighed before filtering the MLSS. After filtering, the filter paper is dried in an oven for 1 hour at 103°C. The suspended solids are calculated using a formula (see laboratory methodology in *Standard Methods for the Examination of Water and Wastewater*, part 2, page 56).

(Source: Gerard Wheeler, Global Facilitation, Inc.)

Figure A-2. Centrifuge detail; to balance the weight applied to the centrifuge place tubes in opposite holders (for example, 1 and 2 or 3 and 4)

Calculate the Spin Ratio

The percentage of compacted solids is used to calculate a centrifuge spin ratio. Figure A–3 shows the result of spinning MLSS in the centrifuge vial. The percentage of activated sludge compacted at the bottom of the centrifuge vial is 2.75%. The concentration of the same MLSS sample tested using Standard Methods was 2,195 mg/L. The spin ratio is calculated by dividing the MLSS concentration by the % spin:

$$2{,}195 \text{ mg/L} / 2.8\% = 798 \text{ mg/L per \% spin}$$

When a spin test is performed, the spin ratio determined by the plant using the above formula is then used to estimate the concentration of TSS present in the sample.

Notes

The accuracy of the spin ratio should be confirmed on a frequent basis to avoid any errors in the actual solids concentration caused by a spin ratio that is not representative. The accuracy of the spin ratio can be checked on a weekly or bi-weekly basis by repeating steps above, and replacing the spin ratio in the data management spreadsheet.

Raw sewage and final effluent samples typically do not contain high enough solids in suspension to accumulate at the bottom of the calibrated centrifuge vial.

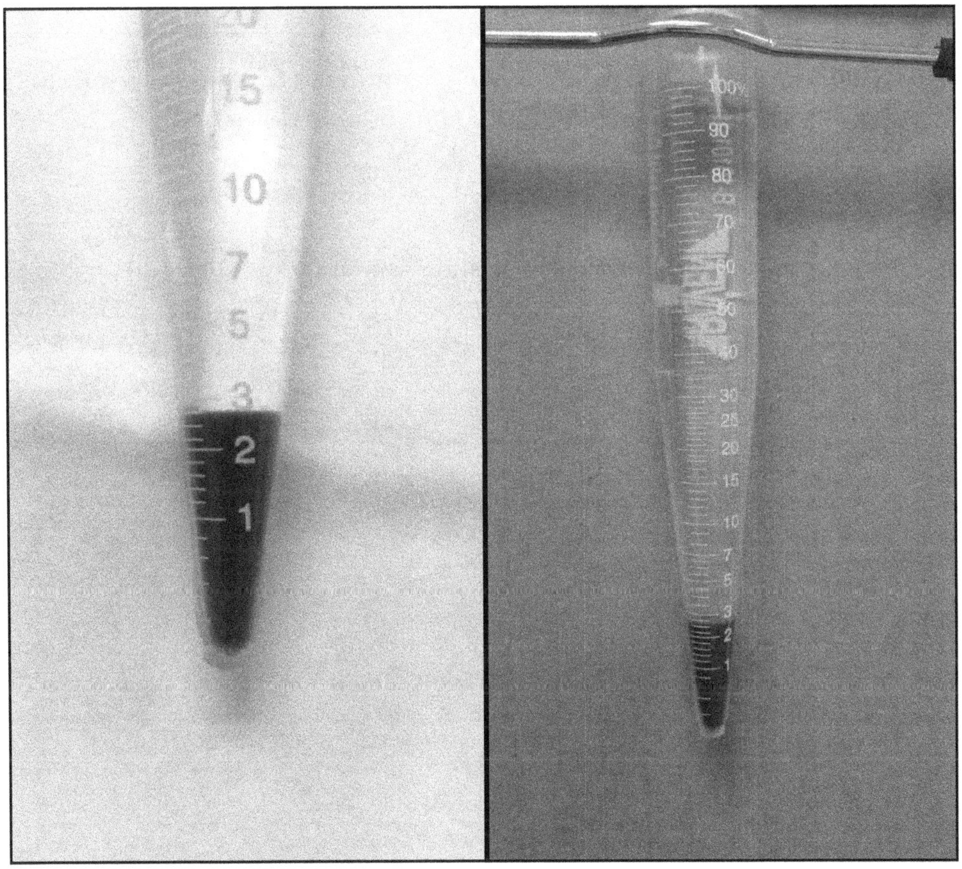

(Source: Gerard Wheeler, Global Facilitation, Inc.)

Figure A-3. Settled solids in the centrifuge tube (left) and the clear water visible above the solids (right)

Therefore the raw sewage and final effluent samples are not suitable for projecting suspended solids using the desktop centrifuge.

To improve accuracy, separate spin ratios can be established for mixed liquor samples from different treatment trains and higher concentration samples (e.g., primary clarifiers and RAS) as each will have different characteristics.

Practical Application of the Spin Ratio

Utility staff can apply the spin ratio to project the concentration of dozens of samples each day without having to spend several hours per day filtering, drying, and weighing glass fiber filter papers. The projected concentration of the samples obtained using the desktop centrifuge are entered daily into a process control spreadsheet that calculates the total sludge mass in the liquid train and scientifically projects the required sludge mass to waste to maintain the desired SRT. This sludge mass control tool spreadsheet is provided to Partnership for Clean Water subscribers. The total sludge mass control concept and SRT are more fully described in chapters 2 and 4 of this guide.

Related Documents

Staff may also reference the related SOP for gravimetrically determining TSS concentration using the procedure outlined in *Standard Methods*.

References

Baird, R.B, Eaton, A.D., Clesceri, L.S, and E.W. Rice (editors). 2012. *Standard Methods for the Examination of Water and Wastewater.* 22nd ed. Washington, DC: American Public Health Association, American Water Works Association, Water Environment Federation.

APPENDIX B

SPECIAL STUDIES

Special studies are designed to help wastewater treatment plant staff more closely evaluate a specific treatment plant issue or perform the activities and data collection needed to evaluate the need for operational modifications. Special studies generally consist of the following steps:

- Problem definition
- Hypothesis
- Approach, including steps taken and data collection requirements
- Duration
- Expected results, including how to evaluate data and how success is defined
- Conclusions
- Implementation

An example special study format is included in this example. If a special study was undertaken during the self-assessment process, utilities should consider including the details of the study, including the outcome, in the self-assessment completion report.

Table B-1. Special study format example

Special study
Special study name:
Hypothesis: Write a hypothesis that is narrow in scope; try to show a definite cause/effect relationship.
Approach: Create a detailed procedure for conducting the study, being sure to involve plant staff in the procedure development.
Duration of study: Define the constraints of the study, which is important because "extra work" is typically involved in completing the study.
Expected results: Describe the expected results, focusing efforts on the interim measurements required and definition of project success.
Conclusions: Document the impact of the study so that it may be used as a training tool for all appropriate staff. Conclusions should be documented even if the special study did not produce the anticipated results.
Implementation: Describe changes to current operating procedures to formalize the mechanism used to improve plant performance.

Source: Renner, R. and B. Hegg. 1997. *Self-Assessment Guide for Water Treatment Plant Optimization.* Denver, CO: AWWA Research Foundation.

APPENDIX C

OPERATOR MATH RESOURCES

Performing mathematical calculations is an important part of an operator's daily functions. A variety of software programs are available, both commercial and utility created, that automate the calculations necessary to determine chemical dilution requirements, chemical dosing requirements, and other related operational parameters. These programs conveniently perform the calculations associated with chemical dosing and plant operation. However, operators should have a basic understanding of operator math and chemical dosing theory. Although mathematical software programs are generally reliable, errors can happen. In these situations, operators should be able to quickly recognize errors and manually set up and perform the calculations necessary to determine chemical dosing rates and other operational parameters. These types of calculations are also a common component of many operator certification exams.

An in-depth discussion of operator math and chemical feed calculations is beyond the scope of this guide, and many operator math and chemistry resources exist from which operators may obtain this information. Wastewater treatment operators are also encouraged to investigate local training options to gain more information about wastewater treatment math, chemistry, and biology. Following is a summary of selected operator math and chemistry resources.

Operator Math and Chemistry Resources for Wastewater Treatment

Girogi, J. 2009. *Math for Wastewater Treatment Operators, Grades 1 & 2*. Denver, CO: AWWA. (Note: Grades 3 and 4 level text is also available.)

Giorgi, J. 2009. *Wastewater Operator Certification Study Guide*. Denver, CO: AWWA.

Singh Sarai, D. 2005. *Basic Chemistry for Water and Wastewater Operators*. Denver, CO: AWWA.

Stubbart, J., W. Lauer, T. McCandless, and P. Olson. 2006. *AWWA Wastewater Operator Field Guide*. Denver, CO: AWWA.

INDEX

NOTE: *f* indicates a figure; *t* indicates a table.

A

D

E

T

U

V

W

www.ingramcontent.com/pod-product-compliance
Lightning Source LLC
LaVergne TN
LVHW080322110825
818359LV00012B/1082
9781625761903